The Open University

Science: A Second Level Course

GEOCHEMISTRY

1 Geochemical Data

2 Composition and Structure of the Earth and its Minerals

3 Geochemical History of the Earth

Prepared by the Course Team

THE OPEN UNIVERSITY PRESS

THE GEOCHEMISTRY COURSE TEAM

J. B. Wright (*Chairman*)
R. S. Thorpe (*Vice-Chairman*)
D. A. Johnson
Joan Mason
Eve Braley-Smith *(Editor)*
Geraldine King *(Course Assistant)*
N. E. Butcher *(Staff Tutor)*
Penny Crompton *(BBC)*
M. Freeman *(BBC)*
D. S. Jackson *(BBC)*

The Open University Press
Walton Hall Bletchley Bucks

First published 1972

Designed by the Media Development Group of the Open University.

Printed in Great Britain by
Staples Printing Group
at St Albans

SBN 335 02131 X

This text forms part of the correspondence element of an Open University Second Level Course. The complete list of units in the course is given at the end of this text.

For general availability of supporting material referred to in this text, please write to the Director of Marketing, The Open University, Walton Hall, Bletchley, Bucks.

Further information on Open University courses may be obtained from the Admissions Office, The Open University, P.O. Box 48, Bletchley, Bucks.

1·1

Contents

Table A

List of Scientific Terms, Concepts and Principles used in Unit 1

Taken as prerequisites				Introduced in this Unit			
Assumed from general knowledge or in HED*	GK or HED	Introduced in S100**	Unit No.	Defined and developed in this Unit	Page No.	Introduced here, but developed in later Unit	Unit No.
atmosphere	GK	absorption spectrum	6	modal analysis	10	chemical composition	
average	HED	acid	9	acid rock	12	and mineral	
climate	GK	alkali	9	basic rock	12	structure	2
histogram	HED	andesite	26	ultrabasic rock	12	chemical features	
mean	HED	atom	6	intermediate rock	12	of soil	6
natural gas	GK	basalt	22	major element	12	sampling of ocean	
ore deposit	GK	bauxite	24, 27	minor element	12	floor	2
soil	GK	biosphere	20	trace element	12	basalt types	2
standard deviation	HED	carbohydrate	10	geochemical		origin of the	
		clay mineral	24	sampling	13	continental crust	5
		colorimeter	7	systematic sampling	13	trace element	
		Conrad discontinuity	22	random sampling	13	geochemistry	2, 6
		continental crust	22	confidence limits	14	isotope ratios	5
		craton	24	geochemical average	15		
		crystal	5	laterite	17		
		diffraction grating	28	bauxite	17		
		electromagnetic		soil horizon	17		
		radiation	2, 28	soil profile	17		
		electron	6	chemical components			
		electron energy shell	6, 7	of soil	17		
		emission spectrum	6	composite sample	22		
		equilibrium	9	average chemical			
		feldspar	24	composition of the			
		geochemistry	24	upper crust	23		
		glacial deposit	25	chemical features of			
		granite	26	biosphere,			
		hydrosphere	24	hydrosphere,			
		igneous rock	24	atmosphere	23		
		intrusion, intrusive	24	precision	24		
		ionization energy	6, 7	accuracy	25		
		isotope	2, 6	sensitivity	25		
		limestone	24	G–1	25		
		line spectrum	6	W–1	25		
		lower continental		gravimetric analysis	26		
		crust	22	spectrochemical			
		mass spectrometer	6	analysis	27		
		mica	24	line spectrum	27		
		mineral	24	internal standard	28		
		neutron	6	calibration graph	29		
		nucleus	6				

* *The Open University (1971) S100* The Handling of Experimental Data, *The Open University Press.*

** *The Open University (1971) S100* Science: A Foundation Course, *The Open University Press.*

Taken as prerequisites				Introduced in this Unit			
Assumed from general knowledge or in HED*	GK or HED	Introduced in S100**	Unit No.	Defined and developed in this Unit	Page No.	Introduced here, but developed in later Unit	Unit No.
		peridotite	22	X-ray fluorescence analysis	30		
		photographic plate	2	analysing crystal	31		
		photon	6	Bragg equation	31		
		prism	6	mass spectrometry	34		
		pyroxene	24	isotope dilution analysis	34		
		quartz	24	isotope tracer	34		
		radiometric dating	2	use of X-ray diffraction in chemical analysis	35		
		rock	24				
		sedimentary rock	24				
		sialic layer	22				
		thin section	24				
		ultraviolet spectrum	2				
		upper continental crust	22				
		valence	8				
		visible spectrum	2				
		wavelength	2				
		weathering	24				
		X-ray diffraction	28				
		X-ray spectrum	2				

Conceptual Diagram

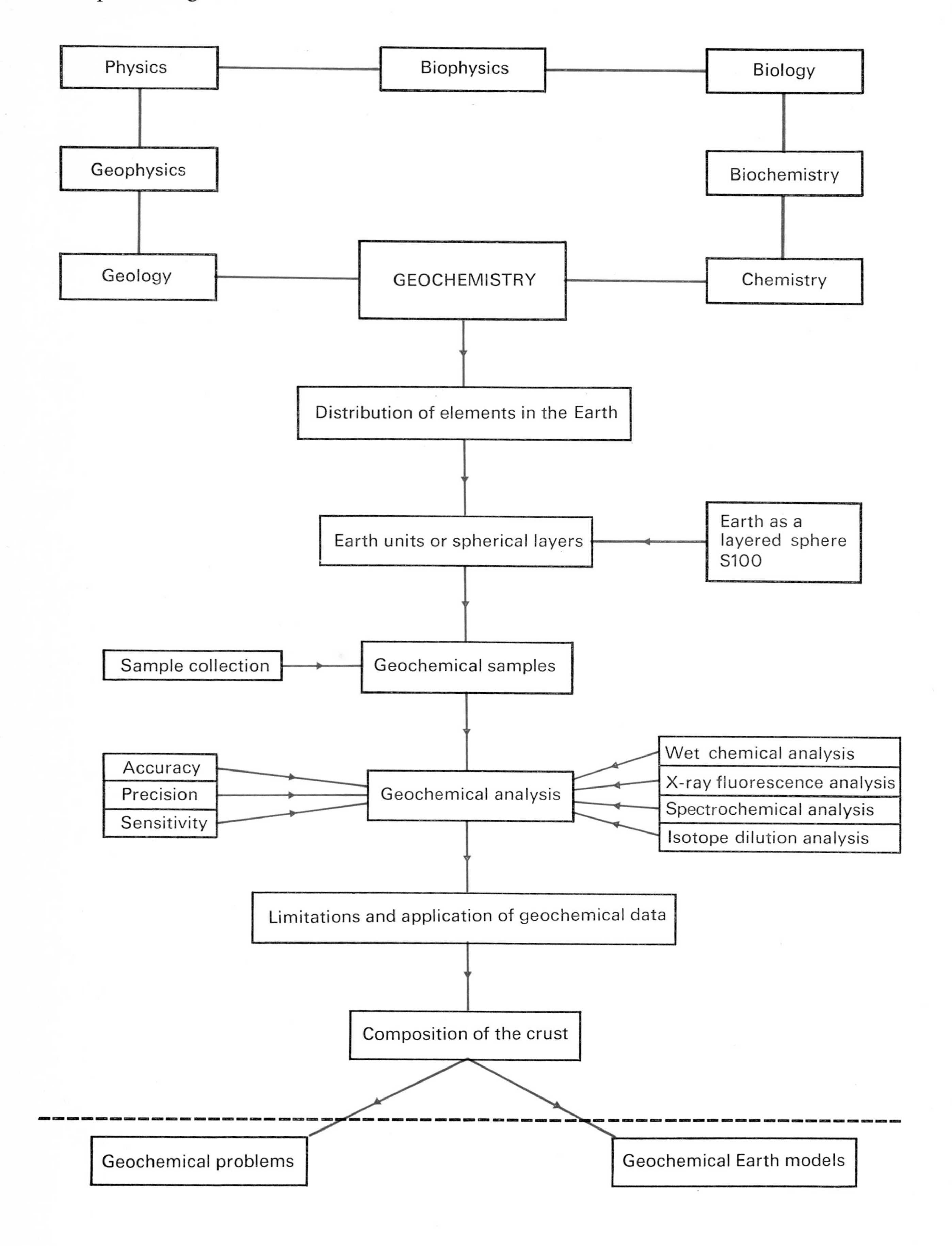

Objectives

When you have completed this Unit you should be able to:

1 Define, or recognize definitions of, or distinguish between true and false statements concerning each of the terms, concepts and principles in the third column of Table A.

2 Critically evaluate geochemical data obtained by sampling, averaging and from chemical analysis, and carry out simple calculations using different geochemical data.

3 Make, or select from a list, qualitative statements concerning the chemical compositions of soils, the hydrosphere, atmosphere, biosphere, and the upper continental crust.

4 List the advantages and disadvantages of methods which have been used to calculate the average composition of the upper continental crust.

5 Explain the principles and outline the procedures used in geochemical analysis by (i) wet chemical, (ii) spectrochemical, (iii) X-ray fluorescence, and (iv) mass spectrometric methods; label a simple diagram of the equipment used for (ii) and (iii); perform calculations on measurements obtained by methods (i), (ii) and (iii).

6 Given data concerning different analytical methods, show how an appropriate method could be used in the solution of a geochemical problem.

1.1 Introduction

In this short Course the main emphasis will be on the methods by which geochemists select and analyse accessible Earth materials, the relationships between the chemical composition and the structure of rocks and minerals, the chemical aspects of the Earth's structural evolution, geochemical reactions and cycles, and the practical applications of geochemistry.

Geochemistry is concerned with the distribution of elements within the different parts of the Earth. This naturally involves consideration of both the physical and chemical properties of the Earth as a whole, and of the chemical processes which take place within and upon it—these correspond to the traditional fields of geology and chemistry. The Course draws heavily on both the Chemistry and Earth Science of the Science Foundation Course (S100), particularly Units 5-12 and 22-27. You will also be required to read sections from the set book *Principles of Geochemistry** by Brian Mason.

1.2 Definition of Geochemistry

In S100 you studied the various aspects of, and the relationships between, different scientific disciplines. You will have realized that the discoveries and achievements of science can often be extended far beyond their original disciplinary boundaries and applied to ever-widening fields. During this century, a number of hybrid subjects such as biochemistry and geophysics have developed, the results of which have had far-reaching consequences. Geochemistry is another such hybrid which is of the greatest practical importance, since it relates to the study of mining, metallurgy, agriculture, soil science, oceanography and geology. Before discussing details of these relationships a brief historical review is important in reaching a proper understanding of what geochemistry is.

Although, in 1821, the famous chemist J. J. Berzelius, called mineralogy 'the chemistry of the Earth's crust', it was in 1838 that the chemist C. F. Schönbein (the discoverer of ozone) used the term 'geochemie' (geochemistry). He also mapped out the programme for this science and wrote: 'we have to study in the greatest detail the properties of all geological formations, we have to find out the relationships of their chemical and physical qualities and their chronological sequence as exactly as possible and we have, at the same time, to compare carefully the products of the chemical forces active at the present time with the organic substance of the past'.

The simplest definitions of geochemistry embrace the 'chemistry of the Earth as a whole', which is the most extensive definition; to appreciate the meaning of geochemistry, we need a more restricted and specific one. In 1924 F. W. Clarke gave such a definition: 'Each rock may be regarded as a chemical system in which . . . chemical changes can be brought about. . . . The study of these changes is the province of geochemistry. To determine what changes are possible, how and when they occur . . . and to note their final results, are the functions of the geochemist.'

Goldschmidt, in 1933, further stated that 'to discover the laws which control the distribution of the individual elements' the geochemist requires 'a comprehensive collection of analytical data on terrestrial material, such as rocks, waters and the atmosphere; he also uses analyses of meteorites, astrophysical data on the composition of other cosmic bodies and geophysical data on the nature of the Earth's interior. Much valuable information has also been derived from the laboratory synthesis of minerals and the investigation of their mode of formation and their stability conditions.' Goldschmidt's definition gives a comprehensive

* *Brian Mason (1966)* The Principles of Geochemistry, *3rd ed., John Wiley (paperback).*

description of the whole subject and in this Course we shall consider geochemistry to be concerned with 'the laws governing the distribution of the chemical elements throughout the Earth.'

Although geochemistry draws on mineralogy, petrology, chemistry and geology, it has existed and developed as a subject in its own right since the time of Schönbein, and has contributed widely to all its related subjects. If you are interested, the history of geochemistry is given in greater detail in *Principles of Geochemistry*, pp. 2–7 (black-page material).

Now examine the Conceptual Diagram on p. 6 which shows how geochemistry is related to other scientific disciplines and also sets out the relationships between the different topics dealt with in this and later Units.

SAQ 1* Which of the following statements are true?

1 Geochemistry is concerned solely with the search for economic deposits of chemicals within the Earth.
2 Geochemistry seeks to explain the distribution of chemical elements within naturally occurring materials.
3 Geochemistry is concerned solely with the application of analytical chemistry to naturally occurring materials.
4 Geochemistry is a mainly theoretical scientific field.
5 Geochemistry has, as its scope, all natural materials in the Solar System.

1.3 Geochemical Data *(Objectives 1, 2)*

The most easily accessible and sampled materials include the atmosphere, hydrosphere, biosphere and the Earth's crust, which together constitute only 0.4% of the Earth's mass. Since this small fraction of the Earth is the most we can sample directly, we are justified in considering it in some detail.

Before looking at any analyses, however, let us examine some of the conventions used by geochemists, for they may seem odd at first. Since rocks and soils are by definition physical aggregates of the natural chemical compounds called minerals†, the most obvious way of expressing their composition is in terms of the relative proportions of minerals they contain. Rock analyses of this kind, determined by simple counting techniques using thin sections (see S100[1])** are known as modal analyses. Approximate modal analyses of a typical granite and basalt are presented in Table 1.

modal analyses

Table 1 Modal analyses of a granite and a basalt

	(% Mineral content by volume)	
	Granite	Basalt
Quartz	30	—
Alkali feldspar†	60	5
Plagioclase feldspar†	5	45
Pyroxene† (augite)†	—	40
Biotite†	4	—
Olivine†	—	5
Iron oxide†	1	5
	100	100

The compositions of these rocks, and of the minerals they comprise, can also be expressed in terms of elemental abundances. The means of chemically analysing rocks and minerals are described in Section 1.7. The results of such analyses are

* *Answers and comments for Self-assessment Questions* (SAQs) *are given at the end of the Unit, p. 38. There will be other questions in the text designed to help you follow specific lines of reasoning; answers to these will be provided immediately following the question so as not to interrupt the subject matter.*

† *Items marked thus are defined in the Glossary, Appendix 2, p. 40.*

** *The list of references corresponding to superior numbers ([1], [2], [3] etc.) in this Main Text will be found in Appendix 1, p. 40.*

usually expressed in terms of element oxides, at least for the major chemical constituents—a convention arising from the fact that oxygen is the most abundant element, at least in the outer parts of the Earth.

It is important to establish the relationship between the mineralogical and the chemical composition of rocks. Minerals are chemical compounds and can therefore be isolated and analysed in the same way as any other compound. Table 2, reproduced from Table 2 on p. 21 in *Understanding the Earth** (*UTE*), presents the approximate chemical composition of some common minerals, expressed as element oxides for the reasons already given above.

Table 2 Approximate chemical compositions of some common minerals (% weight)

	SiO_2	Al_2O_3	$MgO+FeO$	CaO	Na_2O+K_2O
Ferromagnesian Minerals					
olivine	40	—	60	—	—
augite	50	3	23	20	—
hornblende†	40	10	30	12	1–2
biotite	36	15	30	1	10
Plagioclase series					
calcic plagioclase	54	29	—	12	5
sodic plagioclase	68	20	—	—	12
Alkali feldspar	65	18	—	—	17
Quartz	100	—	—	—	—

All the common minerals in Table 2 contain SiO_2 (silica). They are the *silicate* minerals which form the rocks of the outer half of the Earth and which we shall examine in some detail in Unit 2.

Rocks are physical aggregates of minerals. This means that, when they are analysed, the different minerals each contribute, as it were, a proportion of elements to the bulk composition.

Table 3 Chemical compositions of a typical granite and a typical basalt (reproduced from columns 1 and 3 of Table 3 on p. 21 in UTE)

	(% by weight) Granite	Basalt
SiO_2	70.8	49.0
TiO_2	0.4	1.0
Al_2O_3	14.6	18.2
Fe_2O_3	1.6	3.2
FeO	1.8	6.0
MgO	0.9	7.6
CaO	2.0	11.2
Na_2O	3.5	2.6
K_2O	4.2	0.9
	99.8	99.7

SAQ 2 Examine Tables 1–3 and answer the following questions:

(*a*) Why is the SiO_2 content of granite so much higher than that of basalt?

(*b*) Which minerals in the granite and which in the basalt provide iron and magnesium in the analysis?

(*c*) Can you therefore explain why iron and magnesium are much more abundant in basalt than in granite?

(*d*) One constituent in Table 3 is provided by all the minerals in Table 1 (except the iron oxide). Which is it?

Because SiO_2 was formerly regarded as an acidic oxide, the rocks in which it is abundant were defined as acid. Rocks such as basalt in which SiO_2 was balanced by weak basic oxides such as CaO and MgO were termed basic, while rocks

* *I. G. Gass, P. J. Smith and R. C. L. Wilson (ed.) (1971)* Understanding the Earth, *Artemis Press.*

with high MgO content and low SiO_2 content were called ultrabasic. These terms are still used in an arbitrary way for the chemical classification of rocks, as given below:

SiO_2 over 65%	acid
65–52%	intermediate
52–45%	basic
below 45%	ultrabasic

acid
intermediate
basic
ultrabasic

Since the chemical and mineral compositions of rocks are so closely related, these terms can also be applied to rocks simply by looking at their mineral content. Thus a granite is an acid rock because of its high content of alkali feldspar and quartz, which in turn express the high SiO_2 content of the rock. Basalt can be termed basic because it contains a high proportion of minerals rich in CaO, MgO and FeO.

In Table 3 each rock analysis totals nearly 100% when nine oxides are considered. Over 80 other naturally occurring elements are known. Where do these occur? Apart from concentrations in ore deposits (discussed later in the Course), these elements are commonly dispersed in minor (small) or trace (very small) amounts throughout almost all rocks. Many elements can be detected in rocks in abundances greater than a few parts per thousand million, and differences between trace, minor and major elements may be arbitrary. Indeed, elements which are trace elements in some rocks are major components of others.

major elements

minor elements

Trace elements are very important in geochemistry today, partly because of the improved reliability of analytical techniques (Section 1.7), and partly because it has been found that their concentration levels fluctuate more than those of major elements, often by orders of magnitude rather than by factors of two or three. These features of trace elements make them very useful in studying geological problems which are difficult to solve using conventional methods. Moreover, many elements essential to our health enter our bodies in trace quantities (S100[2]) and in a later Unit you will see how the detection of such quantities of metals in soils and natural waters can help in the detection of ore bodies. For the time being, however, we shall confine our attention to the commoner elements.

trace elements

1.4 The Sources of Geochemical Data *(Objectives 1, 2)*

The primary source of much geochemical data consists of rock samples collected at the surface of the Earth, from boreholes or from below the sea. These samples form the basis of our study of all geochemical problems. However, the total weight of samples for a million analyses of rocks might represent a million 1 kg samples, weighing 10^9 g altogether. These would provide the basis of our speculation on the chemical composition of the whole Earth—weighing about 10^{27} g! The choice of samples for analysis is obviously of vital importance. How and where we collect geochemical data forms the basis of this Section.

Large chemical variations occur at the Earth's surface and exposed rocks may be of widely different and chemically unrelated types. Study of a geological map of the British Isles, for example, suggests large variations in rock type.

Although it may not appear so, the Earth's crust has a rather uniform composition. Before we try to find out more about this, there is a more fundamental question that must be considered. The analyses you have looked at so far, and many others that will appear later in the Course, are *averages*. Think carefully about this for a moment. Although an average is advantageous in summarizing a lot of data in a single analysis, it is disadvantageous that we do not know how much variation it conceals, how it was calculated and how reliable it is. Before we progress with the Course, it is important that we consider some of the difficulties in collecting data. Most scientific data is expressed numerically. This is true of many other kinds of data such as sports results, stock market prices and weather reports. In everyday life much importance is attached to figures. News reports giving figures or statistics to illustrate a particular point are often regarded as more reliable than qualitative statements. In particular, trends and comparisons are often illustrated by reference to figures: changes in the frequency of murders, road deaths, examination passes and income levels are demonstrated by using numerical data.

For these and many other cases we are entitled to ask: 'can a change or difference be ascribed mainly to chance, or does it represent a real variation?' Whenever we compare two figures which are averages of many measurements (e.g. summer and winter rainfall) or which are subject to natural or chance fluctuations (e.g. road deaths on Christmas Day) we need (ideally) to know whether the difference between them can be explained by their inherent variability. If a difference can be so explained, then it is not very reliable no matter how large it may be. Furthermore, every measurement is subject to random and systematic errors of which we may often be unaware.

1.4.1 Geochemical sampling

geochemical sampling

Ideally, sampling involves the segregation of a small fraction of a large bulk of material in such a way that the characteristics of the bulk can be estimated by studying those of the sample. (A spoonful of soup will tell you whether it contains enough salt, provided you stirred the soup first!)

Sampling is very important in many fields—you have already read of its importance in biology (S100[3]). Almost all geochemical data is based on chemical analyses of such small fractions, which are often hand specimens of rock taken from very much larger rock bodies. Geochemical theories depend on these samples being representative of the rock masses from which they are taken. It is therefore important to ask how reliable the sampling scheme is for a particular collection of data.

Samples for geochemical analysis can be collected in a variety of ways. The type of sampling plan adopted will depend on the purpose for which the analyses are required and samples collected for one objective may be unsuitable for others. Several plans can be adopted for geochemical sampling; two of the commonest are now considered.

systematic sampling

Systematic sampling refers to the collection of samples in a defined pattern—such as at intersection points on a square grid. Because such a grid can be used to cover a particular area, systematic sampling is the most effective method of collecting samples for the detection, evaluation and interpretation of areal patterns of geochemical variation.

random sampling

Random sampling refers to the independent collection of samples from random geographical locations. Such collection of samples does not usually provide an even coverage of a specific area and it is hence not suitable for areal studies. Random sampling is used in preference to systematic sampling, however, where the samples are to be used for the estimation of average values or of degrees of association between geochemical parameters.

Practical considerations such as the area of rock outcrop may make the implementation of a single sampling plan impossible. In such cases a compromise must be reached between the conflicting factors. An example of this is illustrated in Figure 6b which shows the locations of samples used to estimate the average chemical composition of one part of the Canadian Shield.†

Which sampling plan would you ideally use for this objective?

Because coverage of a particular area is desired, a systematic plan would be ideal. For several reasons it was not possible to use such a plan and the samples actually used reflect as even a cover of the area as was possible.

Although one might think that it is always physically possible to select random samples of rock, it may not be so. In many areas, the rocks are not perfectly exposed and the outcropping rocks may be harder than those covered by soil or stream deposits. Exposed fracture surfaces may be weaker than closed fractures, and such non-random features of exposure may place a chemical bias on the samples collected.

In some cases the practical difficulties place very severe limits on sampling. Some materials—soils and natural gases for example—may show both large spatial variation and a variation dependent on the time of collection. For another important group of rocks—those forming the ocean floor—it is difficult to evaluate the sampling quality, because such rocks are collected mainly by dredge hauls with subsidiary drilling and we do not know how random the collected samples are.

To try to define more closely some of the factors involved in the selection of representative samples, we shall now look at an idealized situation.

Consider a coarsely crystalline metamorphic rock containing 5% of garnet† crystals distributed randomly throughout it. The garnet crystals each contain 40% FeO, and the other minerals in the rock contain no iron.

How much FeO will the whole rock contain?

The whole rock contains 5/100 × 40% = 2% FeO. Since this is an ideal situation we can refer to this as the true value of FeO in the rock.

However, because the garnet is randomly distributed as crystals (let us say of 5 g weight, which means about 1.3 cm in diameter because of the known density of garnet), samples of a given size will not always give the value of 2% FeO on chemical analysis, because different samples, even of the same size, may contain different numbers of garnet crystals.

Even with such large samples, there is a small chance of getting no garnets at all and thus having an analysis with no FeO! However, we are most likely to collect samples with slightly more, or slightly fewer garnets than the overall true figure and this is reflected in the shape of the frequency curve, Figure 1.

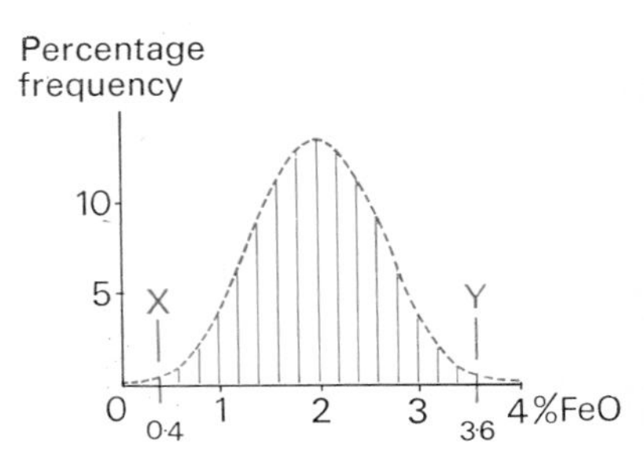

Figure 1 Frequency of FeO determinations in a rock with 5% garnet as randomly scattered grains each weighing 5 g and each containing 40% FeO. This applies to sample weights of 1 kg. Points X and Y represent 99% confidence limits (see Main Text).

What effect will sample size have on our chances of taking a sample with an FeO content close to the mean? This is clearly important since, at one extreme, if we took a sample below 1 cm in diameter, it could well be composed entirely of garnet! To compare the effect of taking samples of different size, we must first look at *confidence limits*.

The confidence limits tell us the chances (probability) of obtaining a result within a certain range of the true value. Take the 99% confidence limit as an example (between X and Y on Fig. 1). If we analysed many samples, 99% of the analyses would fall between those limits; if on the other hand, we analysed only one sample, the chance of it falling within those limits is 99% (i.e. 99 chances out of 100!)

We are now in a position to consider the effect of using samples of different sizes. Figure 2 shows the relationship between the size of the sample and the precision or reproducibility of FeO values at the 99% confidence limit, for our sample of garnetiferous rock.

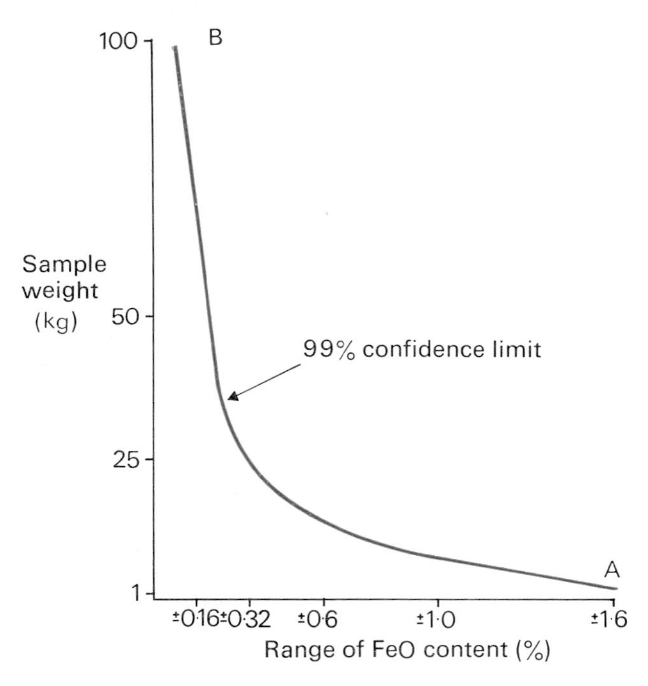

Figure 2 The relationship between sample size and precision of FeO determination (at the 99% confidence limit) in a rock with 5% garnet as grains weighing 5 g each and each containing 40% FeO. For discussion of points A and B see Main Text.

The 99% confidence limits for 1 kg samples provide a precision of no better than ±1.6% FeO (point A on Fig. 2). Thus 99% of many analyses of 1 kg samples would fall within the range 0.4–3.6% FeO between (X and Y on Fig. 1).

Alternatively, if only a single sample were taken, there would be a 99% chance of it falling within these limits.

Could we improve the rather wide range of FeO content at the 99% confidence limit by taking larger samples of the rock?

Figure 2 shows that as the sample size increases, the range of FeO content at the 99% confidence limit falls rapidly at first, but more slowly as the sample size exceeds 25 kg. This represents a very large sample—much larger than most geologists would collect. Even if samples of 100 kg were collected we could still only be 99% certain of having the FeO content to within 0.16% of the true value (point B on Fig. 2).

The large samples are necessary, because in our example we have chosen a coarse-grained rock which contains garnets of unusually large size (~1.3 cm diameter). Most rocks have grain sizes of 1–2 mm or less, so that such large samples are not necessary. Let us now look at a more usual situation.

SAQ 3 Consider a granite containing 25% quartz and 75% alkali feldspar by weight.

(*a*) If the feldspar contains 66% SiO_2 by weight, how much SiO_2 is there in the granite?

(*b*) The confidence limits for determination of SiO_2 in samples of this granite are given in the table below:

Grain size	Sample size	99% confidence limits (% SiO_2)
6·0 mm	1 kg	0·72
1·2 mm	1 kg	0·06
1·2 mm	50 kg	0·009

If you analysed for SiO_2 in 1 kg samples of a granite, having an average grain size of 1·2 mm, and containing the 'true' percentage of SiO_2 worked out in (*a*), within what limits could you be 99% certain that your results would fall (provided your analysis was not faulty!)?

This problem shows that analyses of geological samples have varying significance. A major factor to be considered is grain size as has just been demonstrated by the two preceding examples. In the case of the coarse-grained garnet-bearing rock, samples taken to give FeO analyses reproducible to a significant first decimal place would have to be very large—over 100 kg (Fig. 2). In contrast, to estimate SiO_2 in a finer-grained granite, much smaller samples are needed—in the order of 1 kg—for the equivalent precision.

It is evident that some care needs to be exercised in choosing rock samples for analysis, particularly when a project is designed to determine average compositions, either of parts of the Earth's crust or of a particular rock type.

1.4.2 Geochemical averages

geochemical average

Suppose we have collected and analysed a lot of samples, each representative of its particular rock type—one way of condensing the data would be to calculate averages. This brings us back to an earlier question—how meaningful are average compositions?

This is a difficult question to answer. The averages are computed using widely varying data, obtained by different analysts in different laboratories, from rocks which may be known by different names to different 'experts'. This latter point is important, since rocks often vary gradationally and definitions are not universally agreed.

Let us consider some averages of chemical analyses of basalt, for example. There are several types of basalt, but since they are intergradational, reliable criteria have been sought by which they may be distinguished. Two of the most important groups are the alkaline† and subalkaline† (or tholeiitic), which are common on oceanic islands and continental (circumoceanic) margins respectively. Much effort has been spent on determining criteria to distinguish between basalts from these two environments.

One recent suggestion was that oceanic island basalts have a statistically higher TiO_2 content than circumoceanic ones. To test this hypothesis, a study was made of 857 chemical analyses of basalts from both environments (360 circumoceanic basalts, 497 oceanic island basalts). The results of this study are plotted as histograms in Figure 3.

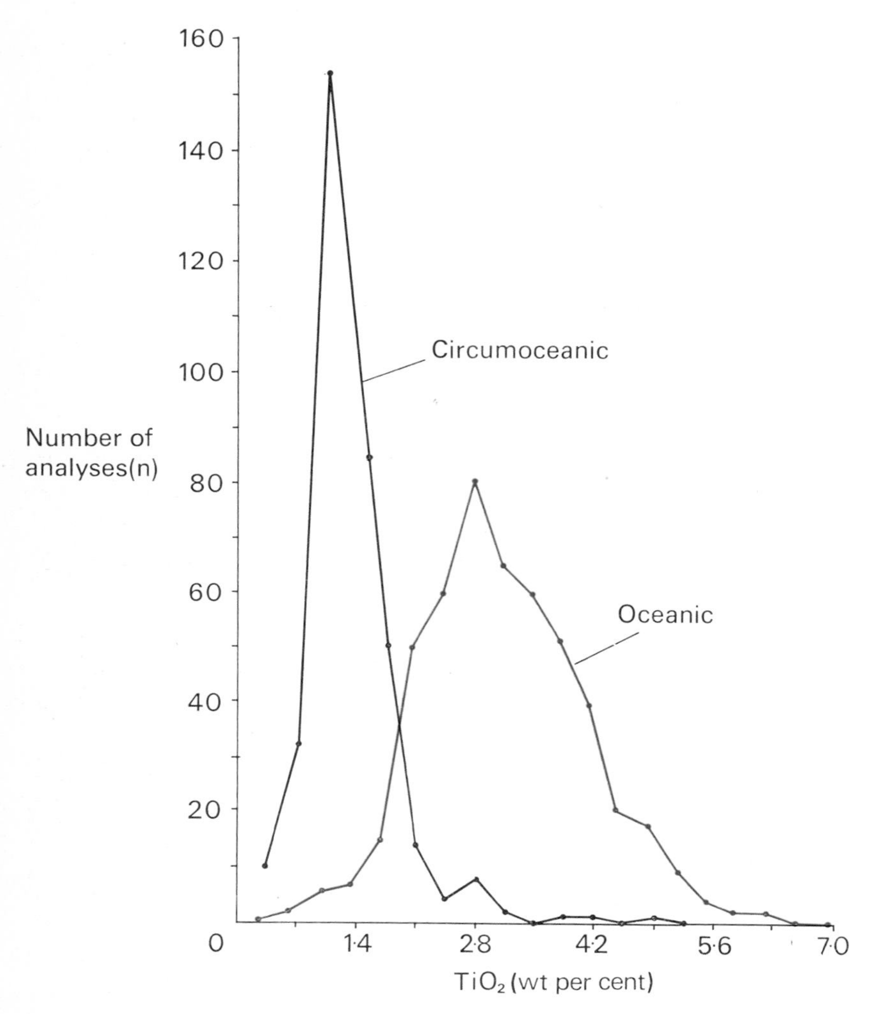

Figure 3 Histogram showing the distribution of TiO_2 in oceanic island and circumoceanic basalts.

The oceanic island basalts had an *average* of 3.06% TiO_2 and the continental margin basalts had an *average* of 1.14% TiO_2. The difference between these figures seems marked. However, these averages conceal a range of values. How certain, therefore, can we be of the original environment of a basalt from its TiO_2 content? To answer this we need to look at the ranges and the frequency of occurrence of TiO_2 contents in the two types of basalt.

SAQ 4

(*a*) Select, from Figure 3, a value of TiO_2 content, intermediate between the two averages, which separates *most* oceanic island basalts from *most* circumoceanic basalts.

(*b*) Work out what proportion of circumoceanic basalts have TiO_2 contents *below* this value.

The number of samples (857 in all) is sufficiently large for us to conclude that the difference in TiO_2 content between these two types of basalt is real. By examining the TiO_2 content of a basalt. therefore, we could be over 90% certain of its type. (Unfortunately, the real picture is more complicated. Basalts from the true oceanic floors resemble circumoceanic rather than ocean island basalts! Resolution of this seeming paradox is beyond the scope of this Course, but is dealt with on pp. 301–3 of *UTE* which you could read as black page material.)

This problem has provided a good example of the use and limitations of averages. Comparison between quantities is possible by using their average values. However, *if we know the range of data that the average conceals, the comparison is very much more meaningful.*

(a)

1.5 Soils *(Objectives 1, 2, 3)*

We consider soils at this point in order to introduce some aspects of geochemical variation, and some of its causes, in a medium with which most of us are familiar. Soils are produced by the physical, chemical and biological breakdown of rocks at the Earth's surface, and are thus the first stage in the production of new sediments and sedimentary rocks, which can be regarded as redistributed soils. Although soils have a negligible mass in the Earth they are obviously vital to our existence since most of our food comes directly or indirectly from the soil.

The chemical composition of soils is very complex; they contain solutions and gases as well as minerals, and many trace elements are present which are as essential as the major elements in ensuring the healthy growth of plants and animals.

Although the composition of soils depends partly on the parent rock, environmental factors, such as climate, are the major cause of differences between soils. A good example of such differences is the development of *laterite* soils in tropical areas; these develop on a variety of rocks (as chemically varied as basalt, granite, and limestone) and are characterized by a high content of Fe and Al oxides. Those strongly enriched with Al oxides and of low Fe content are called *bauxite* and are quarried as the main ore of aluminium.

During the mechanical and chemical breakdown of the mineral particles of rocks by weathering, Na, K, Ca and Mg pass readily into solution. However, Al and Fe are less soluble and react to form insoluble clay minerals† and oxides. These may remain *in situ* or be deposited elsewhere. Tropical areas have intense chemical weathering and may have a low rate of removal of the insoluble minerals.

What factors will favour these conditions?

High rainfall and high temperatures favour chemical weathering reactions. Areas of low relief, common in the tropics, result in a low rate of removal of the insoluble materials. Under these conditions, the ultimate weathering product is developed—laterite (or bauxite in favourable circumstances).

In this brief introduction we have reviewed some of the points made in the section on soil geochemistry, pp. 150-152 in *Principles of Geochemistry*, which you should now read before attempting the following *SAQs*. Use the illustration of soil profiles in Figure 4 as a visual basis for your reading and to help you answer the questions.

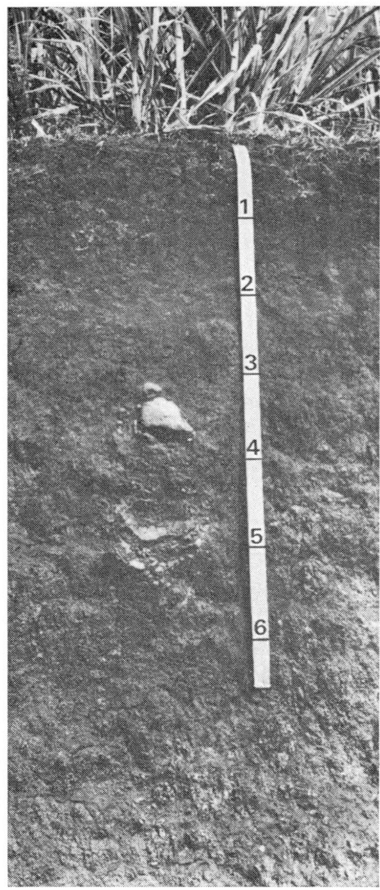

(b)

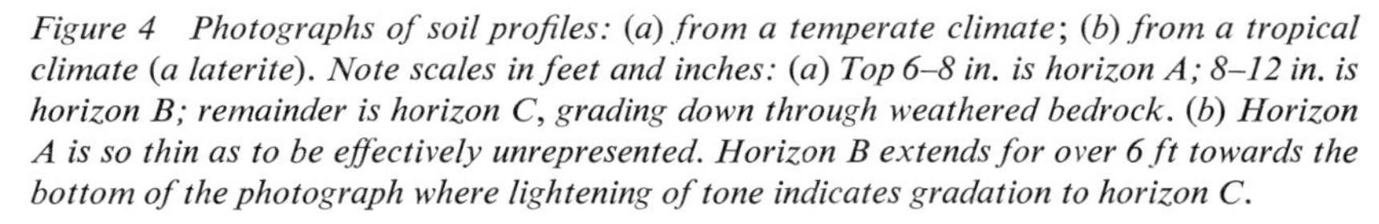
Figure 4 Photographs of soil profiles: (a) from a temperate climate; (b) from a tropical climate (a laterite). Note scales in feet and inches: (a) Top 6–8 in. is horizon A; 8–12 in. is horizon B; remainder is horizon C, grading down through weathered bedrock. (b) Horizon A is so thin as to be effectively unrepresented. Horizon B extends for over 6 ft towards the bottom of the photograph where lightening of tone indicates gradation to horizon C.

SAQ 5 Match the soil horizon in the first list with one or more of the features in the second list.

(i)		
I	A horizon	1 zone of accumulation
II	B horizon	2 characteristically enriched in clay minerals
III	C horizon	3 leaching plays an important part in its development
		4 the parent material for the soil
		5 often enriched in minor and trace elements

(ii) A rock is being weathered and transformed into a soil. Indicate what happens to the elements listed below, during this process.

	Marked decrease	Small decrease	Increase
Ca, Mg, Na, K			
Fe, Al			
Si			

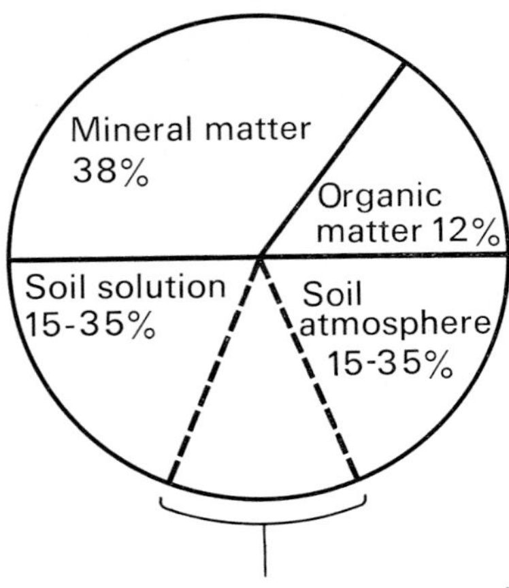

Figure 5 Diagrammatic representation of soil components (expressed as volume %).

The bulk physical composition of a common soil is given in Figure 5 on a volume basis.

Study this Figure and then answer SAQ 6.

SAQ 6 In the table below, indicate which of the substances listed in the left-hand column are contained in the different soil components. Assume that the soil contains sufficient water to form a soil solution with the most soluble of these.

	A Mineral matter	B Organic matter	C Soil atmosphere	D Soil solution
1 oxygen				
2 quartz				
3 feldspar				
4 carbohydrate				
5 clay minerals				
6 sodium chloride				
7 iron and aluminium oxides				

Let us now look at a soil developed at one locality to see how the elements have become distributed between its horizons. As an example, chemical analyses from three horizons of a sandy soil are given in Table 4.

Study these and then answer SAQ 7.

Table 4 Chemical analyses of three soil horizons (weight %)

	Horizon		
	A	B	C
SiO_2	51.84	86.40	92.99
Al_2O_3	2.39	4.92	3.76
Fe_2O_3	0.47	0.79	0.50
TiO_2	0.14	0.17	0.12
CaO	0.86	0.26	0.24
MgO	0.07	0.01	0.02
MnO	0.24	0.02	0.01
K_2O	0.86	1.96	1.66
Na_2O	0.13	0.24	0.36
P_2O_5	0.12	0.06	0.02
SO_3	0.28	trace*	trace
organic matter and ignition loss**	42.49	5.81	0.34
	99.89	100.64	100.02

* *Trace means below 0.01%.*

** *Ignition loss refers to the volatile compounds which burn off when the sample is heated.*

SAQ 7

(*a*) Using Table 4, indicate below which elements (expressed as oxides) appear to be enriched in each horizon relative to the other two. You must think about the level of abundance as well as the variations. For example, TiO_2 is not strongly enriched in any horizon, whereas CaO is enriched in the A horizon.

	Horizon		
	A	B	C
Components enriched			

(*b*) From which rock type in Table 6 (p. 20) is this soil likely to have been derived?

Although soils vary greatly, the data you have just determined reflect general features of soil development in temperate climates such as in Britain. The A horizon is enriched in organic debris and depleted by leaching, especially of alkalis (Na_2O, K_2O) and silica (SiO_2), and to a lesser extent iron and aluminium (Fe_2O_3 and Al_2O_3). This leaching is due to percolation by undersaturated rainwaters. The B horizon is enriched in clay minerals, containing Al_2O_3 and K_2O and hydrated iron and aluminium oxides ($Fe_2O_3.nH_2O$ and $Al_2O_3.nH_2O$), since solutions reach saturation in this zone. The lowest zone (C) may be partly weathered or fresh parent material. These zones are variably developed—we noted earlier that strong leaching (as in the tropics) may leave a soil greatly enriched in Fe and Al oxides. Table 5 gives a chemical analysis of a lateritic soil (Fig. 4b) where this enrichment is great.

Table 5 Chemical analysis of a lateritic soil from Satara, Bombay, India

SiO_2	0.37
Al_2O_3	43.83
Fe_2O_3	26.61
TiO_2	4.45
CaO	0.86
H_2O	23.88
	100.00

A note of caution In *SAQ* 7 we have treated profiles A, B, and C of Table 4 as parts of a closed system, for simplicity. In fact, the organic matter which bulks so large in column A is added from outside the system. In other words, it has *diluted* the mineral constituents of the soil, making the content of inorganic oxides appear much lower than it really is relative to the other two horizons.

For *SAQ* 7 to be truly rigorous, therefore, the organic content should be subtracted from each column in Table 4 and the remainder recalculated to 100%. This modifies your answer in that it emphasizes the enrichments in horizon A but reduces those in horizons B and C. SiO_2 provides a good example: in the recalculation, the figure for horizon A jumps from just under 52% to about 89%, and for horizon B from just over 86% to just under 91%. The progressive decrease is still there, but much less marked.

The generalizations we have made above are not actually invalidated by this; they merely become less obvious. You will appreciate that analyses of soils are more difficult to interpret than any other analyses you are likely to encounter, because they involve both biological *and* inorganic chemical reactions.

1.6 The Average Composition of the Upper Continental Crust

(Objectives 1, 3, 4)

In S100[4] you studied the physical features of the continental crust. One of its most important features is the fact that it can be divided into an upper and a lower part by a poorly defined seismic boundary—the Conrad discontinuity. In this Section, we shall deal with the chemical composition of the crust above the discontinuity—the upper crust. The nature of the lower crust is discussed in Unit 2.

The surface skin of the continental crust is very variable. It consists chiefly of soils, sediments and sedimentary rocks, and supports the biosphere—not to mention artefacts such as roads, cars and buildings! Actually this variation is a result of extreme modification of the Earth's solid continental crust which lies below this surface skin. Were we to determine a weighted average† for all of these surface materials we would get back to the average composition of the upper crust. Tables 4 and 5 illustrated some of the variability in composition of the soil layer. Table 6 covers the compositional *range* of virtually all common rock types encountered at the Earth's surface. In the Table we have presented only analyses for igneous and sedimentary rocks.

Why have we not included metamorphic rocks?

Because metamorphic rocks are simply sediments and igneous rocks which have been recrystallized under the influence of heat and pressure with negligible change in bulk chemical composition.

Table 6 Chemical analyses (averages) of some typical sedimentary and igneous rocks

	1	2	3	4	5	6
SiO_2	95.4	55.1	5.2	70.8	49.0	43.5
TiO_2	0.1	0.9	0.1	0.4	1.0	0.8
Al_2O_3	1.1	16.3	0.8	14.6	18.2	4.0
Fe_2O_3	0.4	4.2	} 0.5	1.6	3.2	2.5
FeO	0.2	1.9		1.8	6.0	9.8
MgO	0.1	2.5	7.9	0.9	7.6	34.0
CaO	1.6	4.7	42.6	2.0	11.2	3.5
Na_2O	0.1	0.7	0.05	3.5	2.6	0.6
K_2O	0.2	3.0	0.3	4.2	0.9	0.2
H_2O	0.3	5.2	0.7	0.3	0.2	0.8
CO_2	1.1	4.0	41.6	trace	trace	trace

1 *Sandstone, a sedimentary rock composed chiefly of quartz grains.*

2 *Shale, a sedimentary rock composed chiefly of clay minerals.*

3 *Limestone, largely calcium carbonate ($CaCO_3$) formed from the shells of animals or by chemical precipitation from sea water.*

4 *Granite* } *from Table 3*
5 *Basalt* }

6 *Peridotite, an ultrabasic plutonic rock composed largely of olivine.*

trace=constituents not present in amounts $>0.1\%$.

Below the surface skin the rocks are more homogeneous, the commonest type being composed chiefly of quartz and feldspar, with some dark minerals, notably mica† and hornblende†. Since silicon (Si) and aluminium (Al) are abundant in these rocks, the crust has been termed the sialic layer (S100[4]). One of the geochemist's aims is to quantify the chemical composition of the sialic layer. In what follows, we attempt this for the upper part, the portion above the Conrad discontinuity.

The calculation of an average upper crust composition illustrates well the problems of sampling outlined in Section 1.4.1. Because the outermost skin, the part most accessible to us, is so variable, you will not be surprised to hear that it is rather difficult to determine what sort of large-scale variations exist between different cross-sections through the underlying crust, and hence to calculate a bulk composition for it.

Read pp. 41–50 in Principles of Geochemistry (*The composition of the crust*). *Do not memorize any figures, but note the different methods for calculating the composition of the continental crust and the problems involved. Use the study comment below as an aid to your reading. When you have finished, answer SAQs 8–14.*

Glossary for pp. 41–50 of Principles of Geochemistry.

PERSILICIC=acid

SUBSILICIC=basic

DIABASE=dolerite (rock similar in composition to basalts but slightly coarser grained).

Study comments for pp. 41–50 of Principles of Geochemistry

1 The averages presented on pp. 42–3 include the exposed rocks of the upper continental crust, as well as the surface skin of soil and sediments.

2 The average presented on p. 44 shows a more basic composition because it includes the oceanic crust. This is no longer strictly a valid procedure since we now regard the oceanic and continental crusts as separate (though obviously related) parts of the Earth's outer carapace. It was once held to be axiomatic that the basaltic ocean crust passes laterally beneath the continents to form the portion below the Conrad discontinuity. This notion is becoming less tenable as more evidence accumulates. We shall consider this topic in more detail in Unit 2.

3 The 'Crustal average' column in Table 3.3 (p. 45) provides elemental abundances for rocks of the exposed continental crust, and therefore refers to the upper continental crust. We shall be talking later about the standard rock types G–1 and W–1.

4 On pp. 48–50, note first the importance of oxygen—to the extent that the lithosphere could also be called the oxysphere; and second, that we are here anticipating a later Unit which will deal with the relationship between geochemistry and ore deposits.

SAQ 8 Look at Table 3.3 in *Principles of Geochemistry*. For the eight elements listed below, express the abundance figures given in column 3 in terms of percentage.

O, Al, Si, Ti, Cr, Fe, Cu, Pb

SAQ 9 What are the inherent difficulties in determining an average composition for the upper continental crust?

SAQ 10 The upper continental crust has a chemical composition:
(i) equivalent to granite;
(ii) equivalent to basalt;
(iii) intermediate between basalt and granite?

SAQ 11 Which is the most abundant oxide constituent in both the continental and the oceanic crust?

SAQ 12 Which is the most abundant element in rocks, and hence in the Earth's crust?

SAQ 13 Why is oxygen more than twice as abundant by volume as it is by weight, while silicon is less than 1/30 as abundant by volume as by weight?

SAQ 14 Why are Na and Ca lower in the average determined by Goldschmidt (p. 43 of *Principles of Geochemistry*) than that determined by Clarke and Washington (p. 42 of *Principles of Geochemistry*)? Tick the appropriate box below.

☐ 1 The Scandinavian crust has a lower content of these elements.
☐ 2 These two elements were leached out during the formation of the clay.
☐ 3 The chemical analyses were not accurate.

The most recent direct studies of the chemical composition of large areas have been made by workers on the Canadian Shield†. This large Precambrian area includes a wide range of rocks. In 1967 Professor Shaw of McMaster University, and his co-workers, sampled several areas of the Canadian Shield (Fig. 6).

composite sample

They collected a total of 8 500 rocks and by combining these into *composite samples* (samples mixed up before analysis in order to obtain averages), analyses of 385 samples for trace elements and 48 samples for major elements were made. From this data it was possible, by careful weighting for rock type abundance as determined from geological maps, to calculate the average composition of each area and hence of the whole Shield.

At about the same time a similar study was carried out by Dr Fahrig and Dr Eade working for the Canadian Geological Survey, using 14 000 hand specimens combined into 180 composite samples. They were able to distinguish between older (Archaean) and younger (Proterozoic) surface crystalline rocks. The contents of the older rocks, some of which may once have been in the lower continental crust, are lower in K_2O, TiO_2, U and Th and higher in Na_2O, Cr and Ni, when compared with the younger rocks. This contrast will be returned to in Unit 2. Meanwhile, the overall chemical composition of the Shield obtained from both studies is given in Table 7.

Table 7 Estimates of the average chemical composition of the surface rocks of the Canadian Shield

NOTE the convention that major elements are expressed as oxides in % by weight, minor elements are expressed as elements in parts per million (ppm). This is universal practice in geochemistry.

	wt %	
	1	2
SiO_2	64.93	65.3
TiO_2	0.52	0.53
Al_2O_3	14.63	15.9
Fe_2O_3	1.36	1.4
FeO	2.75	3.1
MnO	0.07	0.08
MgO	2.24	2.2
CaO	4.12	3.4
Na_2O	3.46	3.9
K_2O	3.10	2.9
P_2O_5	0.15	0.16
H_2O	0.92	0.8
CO_2	1.28	0.2
	ppm	
Cr	99	76
Ni	23	20
Cu	14	*
Zr	400	*
Sr	340	380
Ba	1 070	730
Rb	118	*

1 *Average chemical composition of the Canadian Precambrian Shield (Professor Shaw and co-workers).*

2 *Average chemical composition of the Canadian Precambrian Shield (Fahrig and Eade).*

* *Not determined.*

Compare the averages given in Table 7 first with one another and then with those given on pp. 42–3 in Principles of Geochemistry.

Is there more resemblance *within* or *between* each pair of averages presented? Are the differences between the two sets of data real, or could they simply be ascribed to improved techniques of sampling and analysis?

The differences *between* the two sets of data are more striking than those *within* each set. This is extremely interesting, for the data of Clarke and Washington, and of Goldschmidt, represent markedly different sample coverage (North America and Europe on the one hand, Scandinavia on the other), while those of the Canadian workers represent about the same area. There may well be real differences in the average composition of large segments of upper continental crust, but it is likely that these are comparatively small, and that the differences in the two sets of data are due to improved techniques. This, however, is an opinion, for at present we have no means of verifying such a statement!

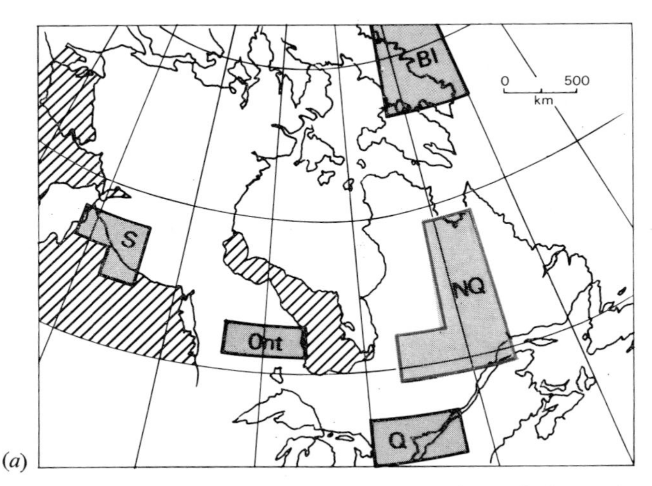

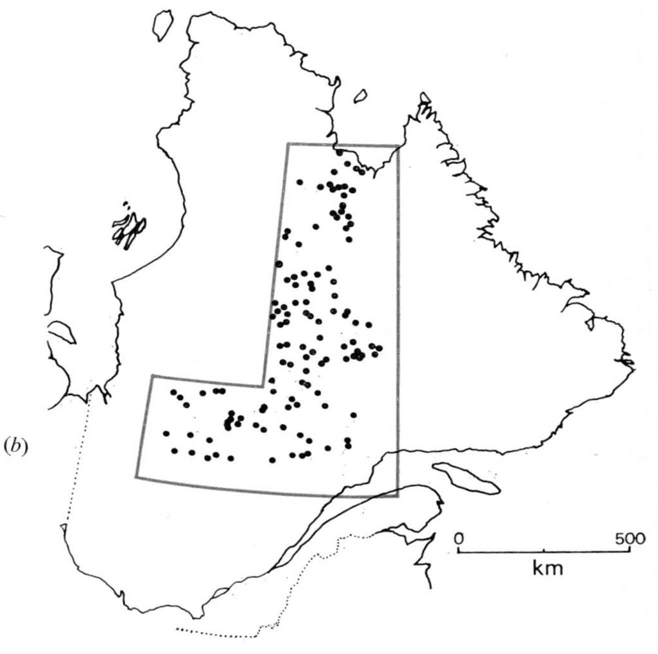

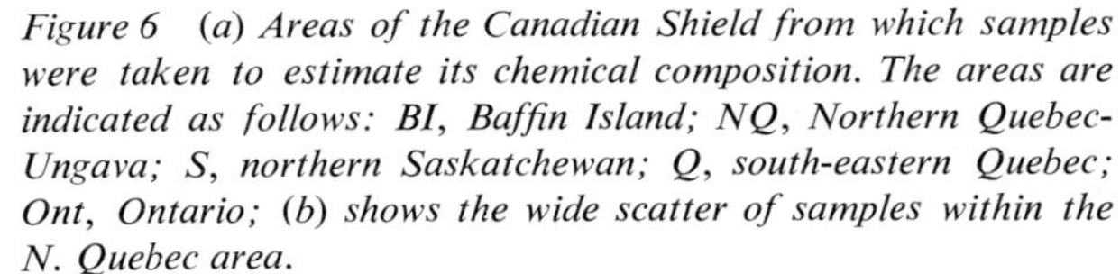
Figure 6 (a) Areas of the Canadian Shield from which samples were taken to estimate its chemical composition. The areas are indicated as follows: BI, Baffin Island; NQ, Northern Quebec-Ungava; S, northern Saskatchewan; Q, south-eastern Quebec; Ont, Ontario; (b) shows the wide scatter of samples within the N. Quebec area.

The Earth's upper continental crust appears to have the overall chemical composition of an intermediate to acid igneous rock. In this composition range, lavas are called andesites† and dacites†, and corresponding intrusive rocks are called diorites† and granodiorites†. Recent estimates of upper continental crust composition provide values close to these of Table 7, some of them slightly more acid still. Such compositions are appropriate to the rock type granodiorite, a common intrusive igneous rock, a microscope thin section of which is shown in Figure 7, and for which compositional data is provided in Table 8.

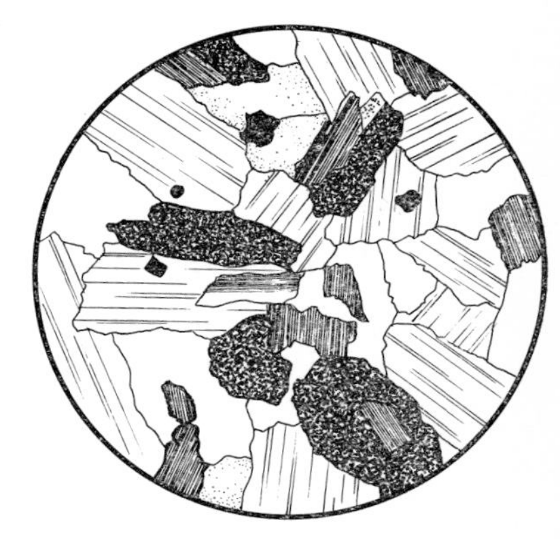
Figure 7 Thin section of granodiorite showing the main mineral components. Approximately half the rock consists of plagioclase feldspar *(clear, with fine discontinuous lines), and about a quarter of* quartz *(clear). The remainder is composed of* alkali feldspar *(weak stipple),* hornblende *(coarse stipple), and* biotite *(closely spaced lines), together with a little* iron oxide *(black). Diameter of field of view is 3 mm.*

Table 8 The chemical and mineral composition of average granodiorite

Chemical composition	
SiO_2	66.88
TiO_2	0.57
Al_2O_3	15.66
Fe_2O_3	1.33
FeO	2.59
MgO	1.57
CaO	3.56
Na_2O	3.84
K_2O	3.07
H_2O	0.65

Mineral composition	
quartz	22
alkali feldspar	18
plagioclase feldspar	50
hornblende	6
mica	2
iron oxides	2

The other outer layers of the Earth The importance of silicon in the crust is due to its chemical properties (S100[5]) and is reflected in the abundance of silicate minerals in the common rocks. The features of these minerals are explored in greater detail in Unit 2. Do not forget that although the dozen or so major elements are of great importance, the remaining 80 or so are widely dispersed and concentrated locally in different parts of the Earth's crust.

chemical features of hydrosphere, atmosphere, biosphere

We should now examine briefly how the major elements are concentrated in the remaining outer layers of our Earth. We can recognize the *hydrosphere*, the *atmosphere* and the *biosphere*. Of these, the first two are fairly distinct, but the *biosphere* pervades the other two and is, of course, also very important in the outermost skin, the soil layer. None the less, these are conveniently recognizable Earth 'shells' and you should get some feeling for their compositional make-up in relation to each other and to the continental crust.

SAQ 15

(*a*) Using *Principles of Geochemistry*, complete the table (in weight per cent to one place of decimals) by referring to the sources indicated below:

1		2		3		4	
O		O	85.9	N_2		O	
Si		H	10.6	O_2		C	
Al		Cl	1.9	Ar		H	
Fe		Na	1.1	C		N	
Mg			——		——	Ca	
Ca		Total		Total			
Na							——
K						Total	
	——						
Total							

1 The crust (Table 3.4, p. 48).

2 The hydrosphere—these values have been inserted for you.

3 The atmosphere (Table 8.1, p. 208). The figure for carbon must be computed from the CO_2 content.

Proceed as follows: Atomic weight of C = } Appendix I in *Principles*
O = } *of Geochemistry*)

molecular weight of CO_2 =

weight % C in CO_2 =

weight % C in average composition of the atmosphere

$$= \frac{460 \times 10^2}{10^6} \times \qquad\qquad =$$

4 The biosphere (use the analysis for man, Table 9.3, p. 231).

(*b*) These analyses illustrate some interesting features.

(i) What is the maximum number of elements needed to account for over 98% of any one sphere?

(ii) Which major element is common to all the 'spheres'?

(iii) In which 'sphere' is carbon concentrated, and why?

1.7 Geochemical Analysis (*Objectives 1, 2, 5, 6*)

In an earlier Section, we reviewed the way in which the size and grain size of rock samples may determine the reliability of geochemical analysis—even before the sample reaches the laboratory! In the laboratory itself other kinds of errors appear, as summarized in S100[6]. There are three contributing factors to error in the laboratory which we shall discuss under the headings: *Precision*, *Accuracy* and *Sensitivity*.

precision

Precision refers to the reproducibility of an analysis figure—the extent to which measurements are spread out about the mean value. Precision is determined not only by the nature of the sample (Section 1.4.1) but also by the skill and/or reliability of the analytical technique and its practitioner. It is always necessary to keep the experimental limitations in mind and although it is often possible to increase precision by expenditure of effort and money, this must be justified by the nature of the problem.

Recording the standard deviation (s) of a set of readings is the best way of expressing the spread of the readings about the mean value. For a large number of readings, s is a measure of how far individual readings are likely to be from the mean. Plotting a large set of readings produces a bell-shaped frequency curve similar to that of Figure 1. Statistical theory shows that about 67% of the readings will lie within $\pm s$ of the mean, 95% within $\pm 2s$, and 99.7% within $\pm 3s$. Standard deviation is therefore a reflection of confidence limits for the analytical technique, in that if 67% of readings fall within $\pm s$ of the mean, those readings represent the 67% confidence limits for analysis by this method.

Acceptable precision, measured by the relative standard deviation or coefficient of variation (C_V), is dependent on the problem under study. Present-day analyses of major element oxides give values for C_V of 0.5–5%, and physical methods for the determination of trace elements may give C_V values of below 0.5%.

SAQ 16 Four analyses of a granite for lead gave results of 9, 10, 7 and 11 ppm.

(*a*) Calculate the mean (x), the standard deviation (s) and the relative standard deviation (or coefficient of variation, C_V, in per cent). The equations for these as given in *The Handling of Experimental Data* (S100[6]) are:

$$\text{mean } \bar{x} = \frac{x_1 + x_2 + x_3 \ldots + x_n}{n} \quad (1)$$

$$\text{standard deviation } s = \sqrt{\frac{(d_1^2 + d_2^2 + \ldots + d_n^2)}{n}} \quad (3)$$

where x_1, x_2 . . . etc. = individual measurements
n = total number of measurements
d = residuals, such that $d_1 = x_1 - \bar{x}$ .... (2)

$$\text{Coefficient of variation} = C_V = \frac{100s}{x}$$

(*b*) What is the expected range of lead values in 95% of granite samples analysed by this method, as deduced from the four values obtained above?

(*c*) Assess the precision of these analyses (C_V).

accuracy

Accuracy is the extent to which analysis figures approach the *true value*. It is difficult to assess because to judge the accuracy of an analysis it is necessary to have reference or standard samples whose composition is known either from previous analysis or from special preparation. These can be analysed at the same time and by the same method as the unknown, thus allowing absolute accuracy to be estimated.

sensitivity

Sensitivity All methods of analysis have a 'lower limit of detection'. This is determined by the point at which a weak signal (e.g. the deflection of a balance needle, the colour of an unknown sample) cannot be distinguished from the background (e.g. zero point of balance, colour of reagents alone). When methods of analysis are described presently, you will see the practical importance of these detection limits.

G–1, W–1

The first serious attempt to assess the precision and accuracy of routine rock analyses was made in 1949 when a large number of identical powdered samples of two rocks: G–1 (a granite from Rhode Island) and W–1 (a dolerite† from Virginia) were prepared by the US Geological Survey and sent to analysts in laboratories throughout the world.

The results from 24 laboratories (64 analyses) were reported and discussed in 1951 by Dr Fairbairn of the US Geological Survey, and surprised many geochemists. For example, in the first 30 analyses of W–1 (up to 1951), SiO_2 values ranged from 51.28% to 53.01%, with a mean of 52.69% and a relative standard deviation (C_V of 0.6%). The precision for trace elements was considerably worse. For example, Sr contents from G–1 (up to 1955) included values of 900, 450, 280, 250, 200 and 120 ppm. Later estimates narrowed the range, so that by 1965 a value of 250 ppm was a reasonable compromise for the content of Sr.

Since any analysis will be both imprecise and inaccurate to some degree, accuracy can only be estimated if these effects can be separated. This was attempted in 1965 by listing three successive attempts (since 1960) to choose 'recommended' (i.e. most accurate) values for the 14 major constituents of G–1 and W–1. However, even for these attempts, agreement to the second decimal place is rare for any constituent. If the second decimal place is meaningful, this means that analytical precision is greater than the accuracy attained at most laboratories. So G–1 and W–1 cannot really be regarded as standards, beyond the first decimal place.

The results of this historic experiment have been far-reaching. Many more geochemical standards are now in circulation and, hopefully, when they become established, the accuracy of future data will be more closely controlled.

You may feel that this survey of errors in sampling and analysis gives a rather pessimistic picture. It is indeed very important to keep these limitations in mind when looking at geochemical data, but we need not be too sceptical of older analyses. In the 1951 report on G–1 and W–1, it was made clear that differen skilled analysts working in the same laboratory and using the same methods could achieve results of high precision. Much important research is done this way. On the other hand, research carried out using analyses from many laboratories, by many different analysts of varying skill, and carried out by different methods, may lead to poorer results.

It is evident from this Section that before embarking on a geochemical project the investigator must specify (as in this Course) the objectives he is seeking. A careful sampling and analytical plan can be devised to minimize the numerous difficulties outlined and enable the geochemist to use his results to the best advantage in satisfying his aims.

SAQ 17 Tick the type of sampling (Section 1.4.1) and quality of the analysis that would be *necessary* to process samples for the problems given.

	Sampling		Analysis	
	Random	Systematic	Accurate and precise	Precise only
1 Samples of soil from Cornwall taken to detect the areas richest in tin				
2 Samples of a large number of minerals analysed for K and Rb in order to determine whether a relationship exists between the contents of these two elements				
3 Samples of metamorphic rocks representative of a particular area in N. Scotland to determine the average potassium content of the crust				
4 Samples of mica from different rock types taken to calculate their average compositions in the different types				

There are many methods of obtaining geochemical analyses, all of which have their advantages and disadvantages, and it is the job of the geochemist to select the most suitable. You have already performed qualitative analyses for elements of geochemical interest in S100[7]. The home experiments that you will carry out in this Course involve qualitative and semi-quantitative analyses of natural materials. The principles used in these analyses are the same as those used in the more sophisticated methods outlined in the next section. Having read about them you should be able to make a choice between alternative methods of solving a geochemical problem.

1.7.1 Wet chemical analysis

Many methods of analysis are based on operations carried out with solutions of rocks—hence they are called 'wet' methods because the rock has first to be dissolved. You have seen a rock taken into solution in S100[8]. The rock was a limestone which dissolved easily in dilute hydrochloric acid. Since most rocks do not dissolve so easily, a stronger acid must be used, for example hydrofluoric acid (HF). Solutions can be analysed in several different ways. Until the last 20–30 years most quantitative data was obtained by *gravimetric* techniques. These involve the precipitation of insoluble compounds from solution, and the calculation of element concentrations from the weight and known compositions of these precipitates. You can see how this is done by answering the following question.

gravimetric analysis

SAQ 18 200 mg (0.2 g) of a rock sample is decomposed and taken into solution. From this the calcium present is precipitated as calcium oxalate $Ca(COO)_2$. This is heated strongly giving 20.8 mg of calcium oxide (CaO). What is the percentage of CaO in the rock? Which rock type in Table 6 has a similar content of CaO?

These older gravimetric methods are rather slow. Since 1950 new schemes of wet chemical analysis have been developed which do not involve precipitation and weighing operations. Many of these are based on the addition of complexing reagents to the rock solution under closely controlled conditions, to form coloured compounds with specific elements. The intensity of the resultant colour is proportional to the concentration of the element. Intensities are measured with a colorimeter (S100[9]) or a spectrophotometer†. You will use a colorimetric method for the partial analysis of a soil sample in your home experiment. These colorimetric methods are more rapid than the gravimetric methods, but they require moderate chemical manipulation and may seem tedious compared with the purely physical methods described later. It is unlikely that wet chemical methods will ever be entirely superseded, however, because although they are lengthy, they are capable of producing very accurate analyses which can be used as standards for physical methods. Physical methods, none the less, have many advantages over the purely chemical techniques and today many laboratories have a choice of methods available. Several of the most important are compared in the following Sections.

1.7.2 Spectrochemical analysis

spectrochemical analysis

You have studied the broad features of atomic structure in S100[10]. Each atom consists of a nucleus surrounded by shells of electrons, as illustrated diagrammatically in Figure 8.

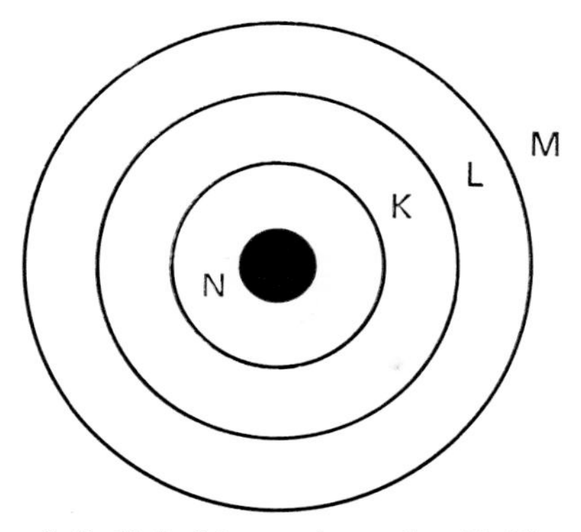

Figure 8 Electron shells K, L, M around a nucleus N. Electrons in the K shell are more tightly bound than those in the M shell.

When atoms are excited by strong heating, the more loosely held electrons in the outer shell (M in Fig. 8), the valence electrons, move to higher levels still further from the nucleus. As the electrons return to their original levels, their excess energy is emitted as visible or ultraviolet radiation of specific and characteristic wavelengths.

line spectrum

The resulting *optical emission spectrum* consists of a series of lines (S100[11]) and is therefore called a *line spectrum*. Each element has a characteristic line spectrum, as you know, having used this principle to detect Li and Ca in an unknown sample (S100[7]). *Emission spectrographic analysis* is useful for elements which are readily excited because they have relatively low ionization potentials (e.g. the alkali metals), but is difficult to use in the detection of elements that are less easily excited (e.g. zinc).

The equipment used is shown in Figure 9. The analysis begins with the excitation of the atoms in the sample by strong heating. The emitted energy consists of electromagnetic radiation in the visible and ultraviolet parts of the spectrum (wavelength 2300–9000 Å). This radiation is refracted through a prism which separates out the different wavelengths present, producing a line spectrum which is recorded on a photographic plate.

Each element has emission lines of specific and characteristic wavelength which correspond to different electronic transitions within atoms of the same element. The *intensity* of each line is related to the *concentration* of the element being excited.

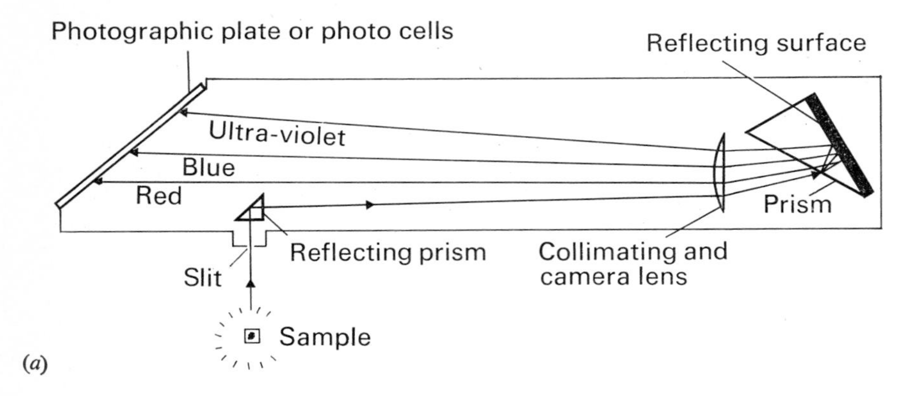

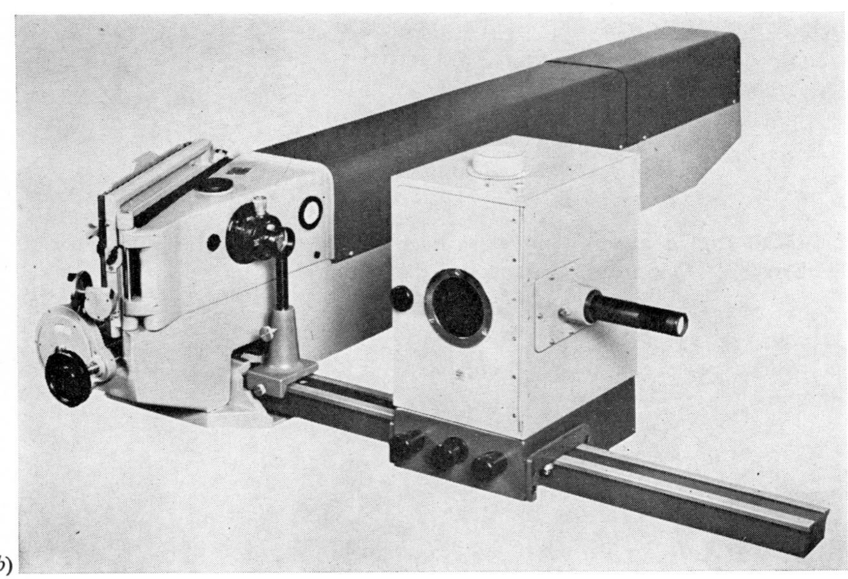

Figure 9 The emission spectrograph: (a) diagram of the optical parts in a prism spectrograph, (b) photograph of an emission spectrograph.

SAQ 19 Look at Figure 10. By looking at the K line of wavelength 6707 Å and the Rb line of wavelength 7947 Å, place the three materials analysed (granite G–1, dolerite W–1, and a mica) in order of decreasing K and Rb content. (You will get the same answer whichever lines you use—these lines are slightly clearer.)

		K	Rb
Highest content	1		
	2		
Lowest content	3		

The intensities of the different emission lines are also measured on the photographic plate. To relate these line intensities accurately to concentration, a conveniently chosen element, or compound of known composition, is mixed in accurately known amounts with the sample. This is called an *internal standard*, and the ratio of the intensity of a specific element line *relative* to an element line in the standard is measured. Because differences in the excitation conditions of each analysis will affect the sample and internal standard alike, this procedure ensures that such differences do not affect the precision of the analysis.

internal standard

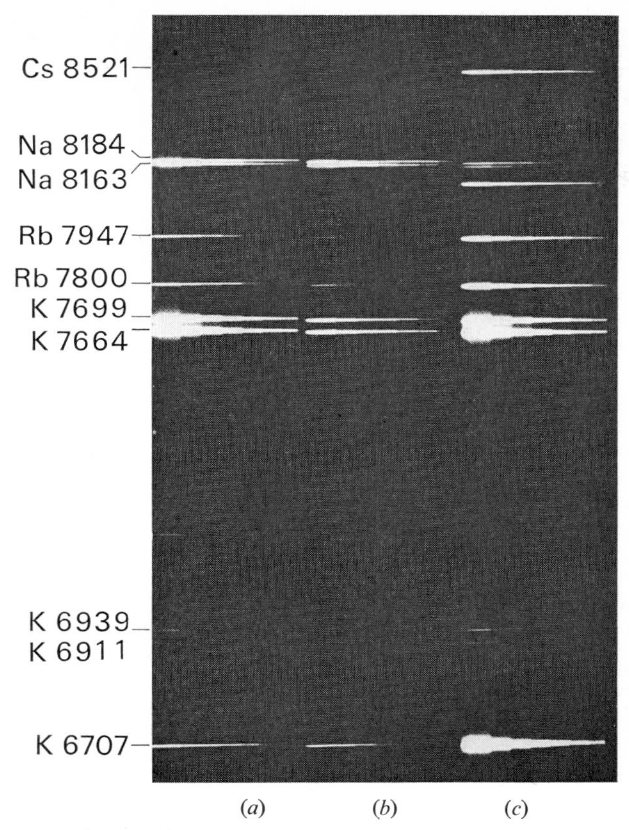

Figure 10 Photographic plate showing alkali metal emission spectra of (a) Westerly granite, Rhode Island (G–1); (b) Centerville dolerite, Virginia (W–1); (c) Mica (lepidolite).

The calculation of the element concentration from its line intensity is done using a *calibration graph*. This is constructed by plotting the line intensities (measured relative to the internal standard) of a set of samples with known concentrations. An example of such a calibration curve for Cr, using palladium (Pd) as an internal standard, is shown in Figure 11.

calibration graph

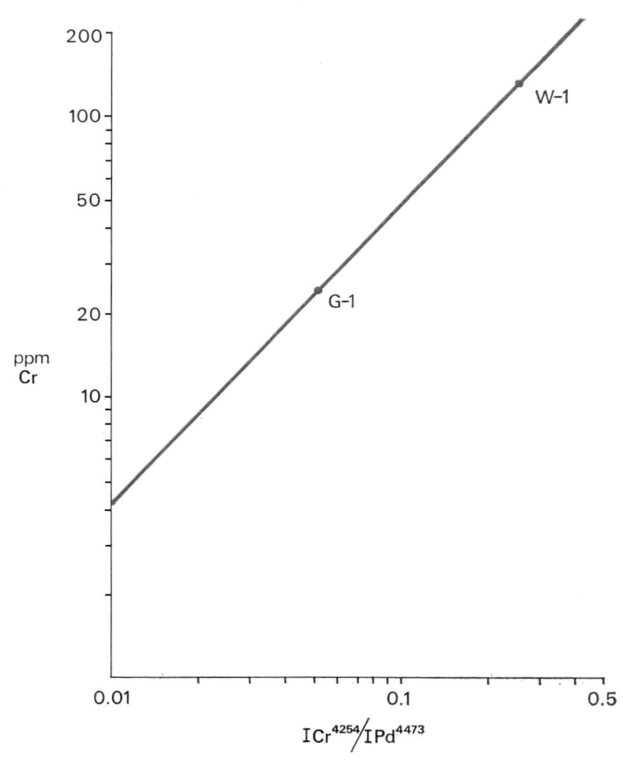

Figure 11 Calibration graph for the determination of Cr in igneous rocks using Pd as an internal standard. The line has been determined by using analytical results from G–1 and W–1.

This method of analysis can produce a vast array of data with relative ease and rapidity—careful organization of results is thus essential. The accuracy and precision depend on the quality of the internal standard and the calibration graph used, and because these may vary greatly, the method can justifiably be used in a qualitative way for some investigations (*see SAQ 19*).

SAQ 20 During a determination of chromium by optical emission analysis, palladium (Pd) is used as an internal standard. Relative intensities were recorded for two emission lines: a Cr emission line with a wavelength of 4254 Å and a Pd emission line of wavelength 4473 Å. The rock analysed was a serpentinite† of unknown origin. Such rocks may originate by metamorphism of ultrabasic igneous rocks or dolomitic limestones†. The average Cr content of these two rock types is given below:

	Average Cr content (ppm)
Dolomitic limestone	12
Ultrabasic igneous rock	1600

Calculate the intensity of the Cr line relative to that of the internal standard from the data tabulated below. Use the calibration graph (Fig. 11) to determine the Cr content of the serpentinite in ppm.

	Intensity Cr^{4254}	Intensity Pd^{4473}	$\frac{I\ Cr^{4254}}{I\ Pd^{4473}}$
sample	12	600	

Which type of rock has been metamorphosed to form the serpentinite sample you have studied?

1.7.3 X-ray fluorescence spectrography

Although the principles of this method have been known since the 1920s, only recently have technical problems been sufficiently overcome to construct automatic spectrometers. These are now very popular, being used in many universities, and are rapid, precise and flexible in operation.

X-ray fluorescence analysis

The principle is similar to optical emission spectrographic analysis but, in this case, the atoms are excited by a primary X-ray beam rather than by heating. The greater energy input causes electrons in energy shells nearer the nucleus to move to higher energy levels, for example from K to L in Figure 12.

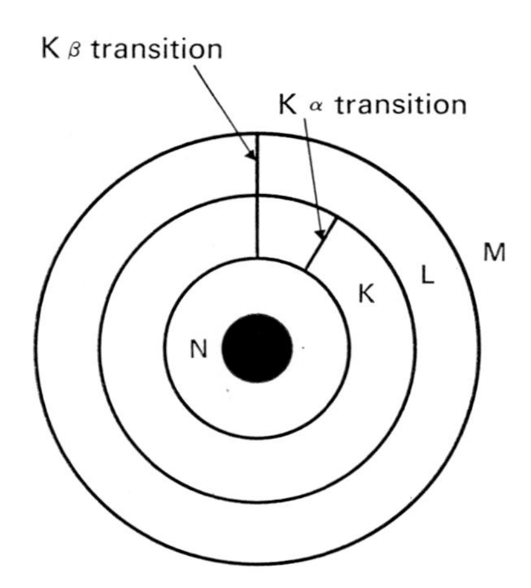

Figure 12 Transitions between electron shells K, L, M around a nucleus N.

The excited atoms revert to their lower energy states by the emission of electromagnetic radiation, but in this case it is of X-ray wavelength (0.1–50 Å). Note two transitions on Figure 12 one between the K and L shells and one between the K and M shells. These are called the Kα and Kβ transitions, respectively.

Each of these transitions emits X-rays of different specific wavelength which, in turn, characterize the different elements. From the intensities of these X-rays the concentrations of elements in the samples can be calculated. A photograph of an X-ray spectrometer is shown in Figure 13a and illustrated diagrammatically in Figure 13b.

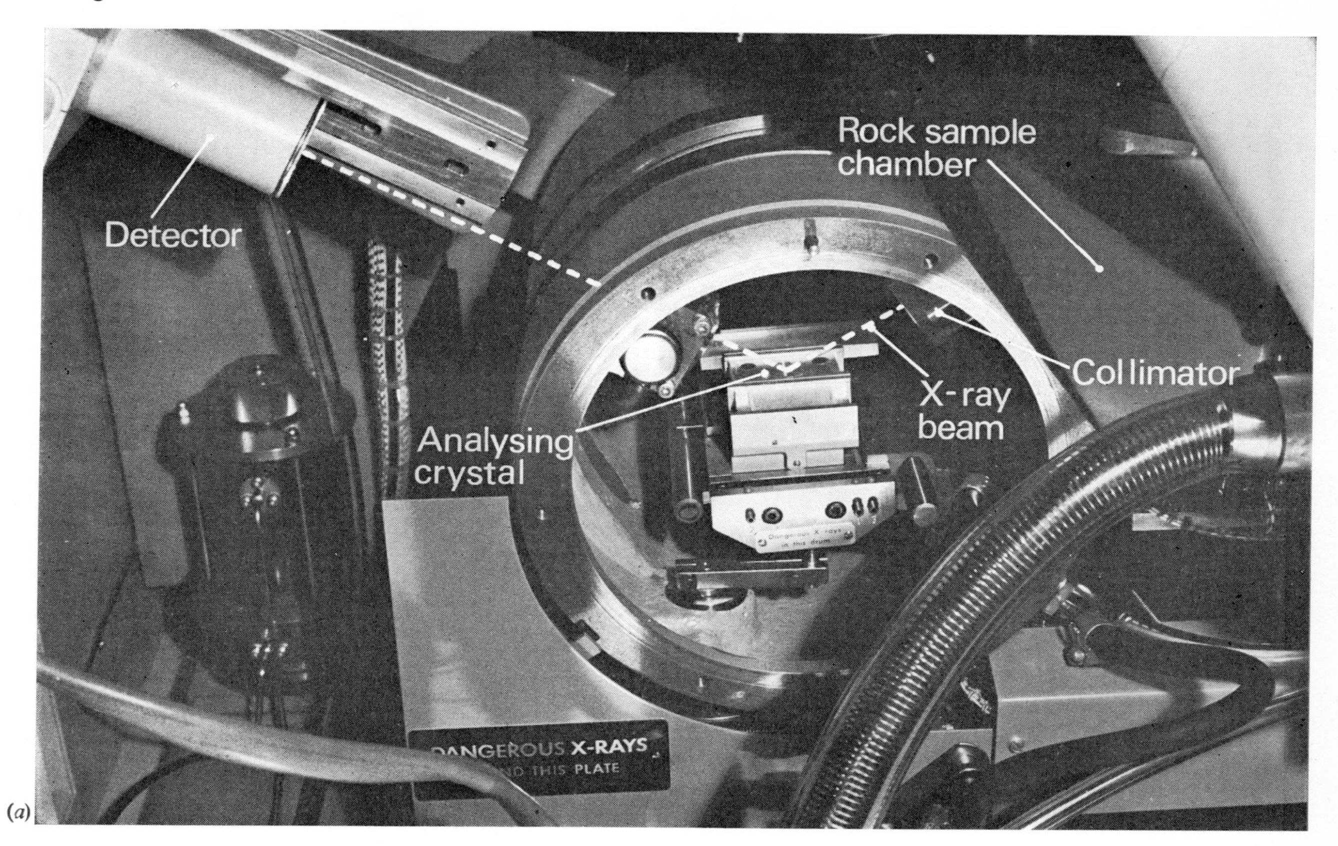

(a)

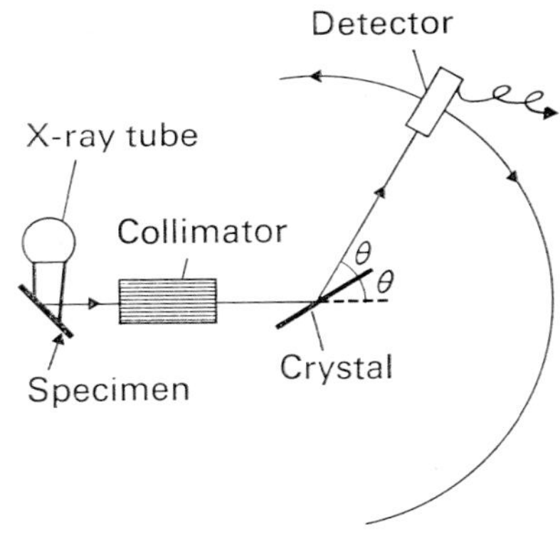

(b)

Figure 13 (a) Photograph of the working parts of an X-ray spectrometer; (b) diagram of essential working parts of an X-ray spectrometer.

X-rays produced by high voltages in the generator are directed on to the sample, exciting the elements in it. Since many elements are present, the secondary (or fluorescent) X-rays so produced have many different wavelengths. These must be separated out.

Think back to the way in which different wavelengths were separated out in optical emission analysis.

Could we use a prism here too?

No, because the X-rays have wavelengths so small that a grating or prism will not separate them. Instead we use an *analysing crystal* which diffracts the X-rays according to the *Bragg equation*

analysing crystal

$$n\lambda = 2d \sin \theta$$

Bragg equation

where d is the spacing of atomic planes in the crystal, λ is the diffracted X-ray wavelength, n the order of the diffraction (S100, Unit 28), and θ is the angle of incidence of the X-rays on to the atomic planes of the crystal (Fig. 14).

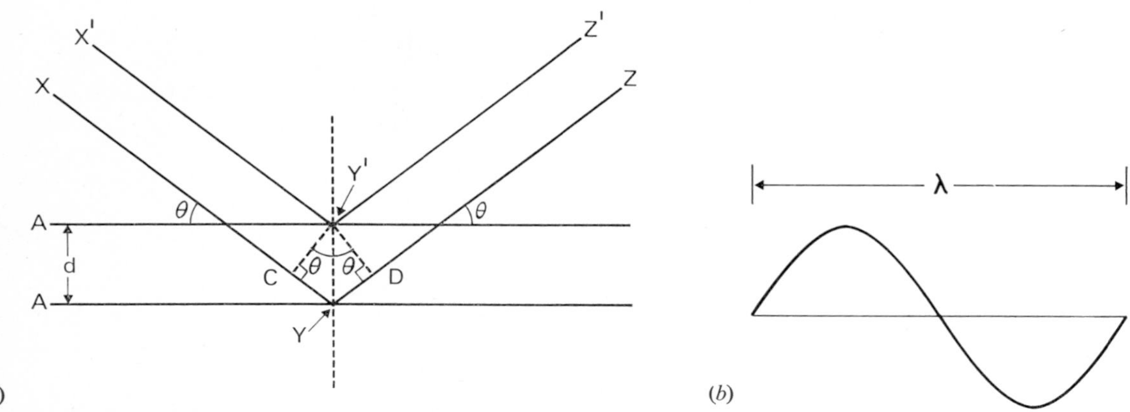

Figure 14 (*a*) *The reflection of a beam of X-rays from different planes in a crystal. Two crystal planes A are shown, the separation between them being designated by* d. *Reflection (diffraction) occurs when the angle of incidence* θ *is such that the distance CYD is equal to a whole number of X-ray wavelengths, as in* (*b*). *At this point* $n\lambda = 2d \sin \theta$. *A derivation of this equation (the Bragg equation) is given in Appendix 3 (black).*

Because we know d for the crystal very accurately and we can set θ to any desired value, radiation of any wavelength can be diffracted into the detector (Fig. 13). The intensity of radiation is measured relative to an internal standard, and is proportional to the concentration of the element being excited.

SAQ 21 For the analysis of silicon, the Si Kα X-ray emission line ($\lambda = 7.13$Å) is used with an analysing crystal having a d spacing of 4.37Å. By using the Bragg equation $n\lambda = 2d \sin \theta$, calculate the angle θ at which the analysing crystal must be set (see Fig. 13b) to diffract the *first-order* diffraction line ($n = 1$) into the detector.

Calibration graphs relating X-ray intensity to concentration can most easily be prepared in an empirical way by using a number of samples containing already known amounts of the elements concerned. The ratios of the X-ray

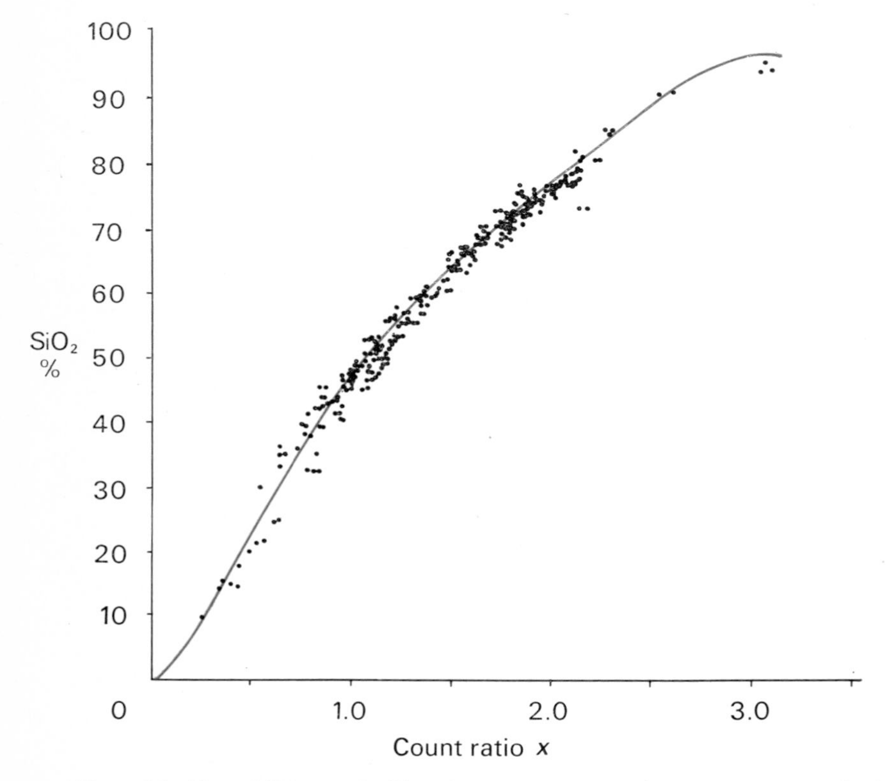

Figure 15 Plot of SiO_2 *weight % against count ratio x, where x is intensity of SiK*α *ratioed against SiK*α *for the internal standard. 621 rocks of all types are represented.*

intensity values of the samples to an internal standard are measured, the internal standard being repeatedly analysed throughout the calibration procedure to allow for instrumental fluctuation. The same standard must, of course, be used with any unknown sample for which the calibration graph is to be used.

Figure 15 is a calibration graph for SiO_2 concentration in about 600 previously analysed rocks. Equations derived from such graphs can also be used to determine element concentrations.

SAQ 22 Analysis of two basalt samples for titanium by an X-ray fluorescence method gave the following data.

	X-ray intensity (counts/unit time)		
	Internal standard	Unknown basalt	Count ratio
Basalt I	996	1992	
Basalt II	1010	4040	

(*a*) Calculate the count ratio for the unknown basalts relative to the standard. Note that the titanium contents of the rocks will be expressed as TiO_2, although the count data was determined from the intensity of the Ti Kα peak.

(*b*) The contents of TiO_2 in the rocks can be obtained from the equation

$$TiO_2\ (\%) = 0.77 \times \text{count ratio} - 0.21$$

which is derived from the calibration graph, Figure 16.
Use both the equation and the graph to calculate the TiO_2 contents of the two basalts.

(*c*) Referring back to *SAQ 4*, would you classify these as ocean island or circumoceanic basalts?

(*d*) Assuming the spread of points on either side of the line in Figure 16 to represent approximately the 99.7% confidence limit, estimate the standard deviation (s) for these two basalt analyses (99.7% confidence limit at $\pm 3s$). Is this large enough to raise a doubt about the answer to (c)?

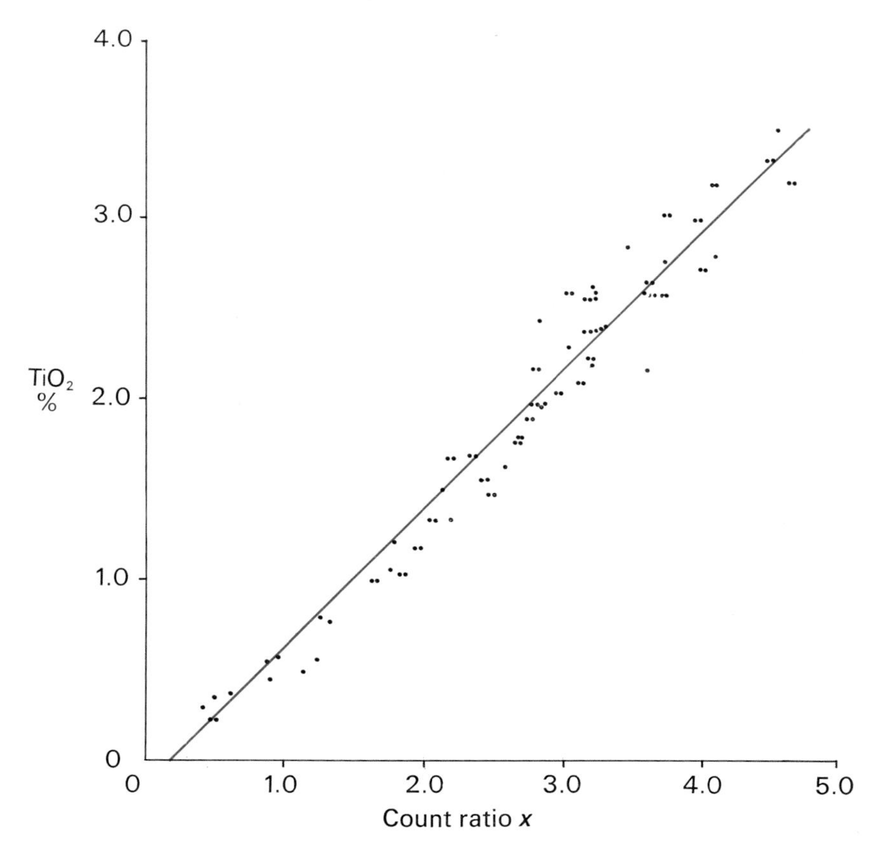

Figure 16 Plot of TiO_2 weight per cent against count ratio x, where x is intensity of TiKα ratioed against TiKα for the internal standard. 128 basic rocks are represented.

1.7.4 Mass spectrometry

mass spectrometry

The methods of analysis we have looked at so far have depended on the different chemical properties of elements.

> Can you remember what atomic property determines the chemical behaviour of an element?

The number of electrons in the outer shells. There is, however, another fundamental atomic property which can be used to distinguish between atoms. It is the number of neutrons in the nucleus. Some elements have naturally occurring isotopes and this is of great geochemical importance. Isotopes and isotopic ratios are studied by using the mass spectrometer—the operation of this instrument was discussed in S100[12]. The most important application of this technique in geochemistry is in the radiometric dating of rocks, as discussed in S100, Unit 2, but it is also possible to use mass-spectrometric methods for very accurate quantitative analysis of elements. Several isotope ratios determined by mass spectrometry have geological importance beyond that of age dating because they give us information about the ultimate origin of rocks and about their temperature of formation. Some of these applications are discussed in greater detail in Unit 5.

The use of the mass spectrometer for chemical analysis is limited to those elements which have more than one isotope. This excludes monoisotopic elements such as F, Na, Al, P, Sc, Mn, Co, As, Y, Au and Bi.

isotope dilution analysis

Because mass-spectrometric measurements of isotope ratios are very precise, the technique has been modified to determine *absolute concentrations* of elements in very low abundance in rocks and minerals. This analytical technique is called *isotope dilution*. It is a highly specific analytical method which enables many trace elements to be determined with high precision, sensitivity and absolute accuracy and can, of course, be used only for the estimation of elements which naturally occur in more than one isotopic form. One example is lithium, the naturally occurring form of which is a mixture of ^{7}Li and ^{6}Li in which $^{7}Li/^{6}Li = 12.4$. A sample containing this lithium would be mixed (diluted) with a measured amount of lithium having a different and exactly known isotopic ratio. The lithium added is known as a *tracer;* in this case a lithium tracer with $^{7}Li/^{6}Li = 0.004$ could be used.

isotope tracer

By measuring the $^{7}Li/^{6}Li$ ratios of the normal and the diluted sample (the amounts of which are both known), it is possible to determine the absolute abundance of lithium. Sensitivity can be very high and may reach 10^{-6}–10^{-12} g. Some of the quantities which have been measured include the following:

Element	*Level of concentration*
U in stony meteorites	0.0105 ppm
U in feldspar	0.22 ± 0.03 ppm
Ne in stony meteorites	$1–2 \times 10^{-8}$ ml/g

High precision is possible; C_v can be as good as 1–3% at levels below 1 ppm. If errors of chemical preparation and contamination are avoided, systematic errors can only arise if the isotope ratio is incorrectly measured or if the quantity of tracer added is not accurately known. An absolute accuracy of ±1–5% should be possible for solids, with better results for gases.

This method of element abundance determination has found wide application in the determination of elements present in low abundances, e.g. Ar, Cs, He, Ne, K, Rb, Sr, and U in meteorites and ultrabasic rocks. Isotope dilution is much slower to perform than the two methods described previously, but it is free from the systematic errors which may be present in the others.

1.7.5 Comparison of physical methods of analysis

We have now looked at three important methods of geochemical analysis and indicated some of the problems to which they can be applied. These three methods are ideally suited to particular problems and before a method is selected for the study of a problem, all its aspects should be carefully considered.

Mass spectrometry is the only method applicable to the determination of isotope ratios in rocks and, in general, the isotope dilution method is the most accurate for element analysis. However, because the isotope dilution method is also relatively lengthy, it is usually used only for the analysis of small numbers of elements present in very low concentrations.

Emission and X-ray spectrographic methods are more rapid and flexible and are often used in the analysis of large numbers of samples for a wide range of elements. Choice between the two may be determined by their different detection limits (Table 9).

Table 9 Detection limits for selected elements using different physical methods of analysis (ppm)

	Emission spectrography	X-ray fluorescent spectrography	Isotope dilution (mass spectrometry)
Al	<5	100	*
B	10	*	**
Be	10	*	**
Ce	100	<5	**
F	100	2 000	*
Li	<5	*	**
Mg	<5	200	**
Nb	30	<5	**
Th	100	<5	**
U	100	<5	**
Zn	100	<5	**

* *Not possible to analyse.*
** *Extremely low detection limit—often below 1 ppm.*

Table 9 shows that for some elements, notably those of low atomic number (below 9), it is impossible to use X-ray fluorescence because the characteristic X-rays for these elements are of insufficient intensity. Although the limits in this table are, to a certain extent, dependent on the analytical conditions, it is these limits which may determine the use of a particular method for a particular problem.

However, subject to these limitations, X-ray fluorescence is superior to emission spectrographic analysis because many detection limits of important elements are low and because automated X-ray spectrometers are now being made. These make high productivity possible—about 40 rocks per week can be analysed for 23 elements. X-ray fluorescence is being increasingly used for projects where many analyses are needed, such as the detection of chemical trends over wide areas and to determine the compositions of large areas of crust.

1.7.6 X-ray diffraction

use of X-ray diffraction

X-ray diffraction is a most important technique in many aspects of geology and geochemistry, but it is not dealt with in detail at this point because it does not yield chemical analytical data directly. It has, however, been used to determine the atomic structures of minerals since the 1920s. These structures are of great geochemical importance since they have formed the basis of our knowledge of the relationship between the structure and composition of the silicate minerals, which have been of the greatest importance in predicting element behaviour, as outlined in Unit 2.

In Section 1.7.3 we described how a crystal could be used to separate out X-ray wavelengths. This was possible because of the relationship between the d spacing of the analysing crystal and the wavelength of the diffracted X-rays, the Bragg relationship: $n\lambda = 2d \sin \theta$ (see Fig. 14 and Appendix 3, black page).

For X-ray diffraction this principle is applied in reverse. A beam of X-rays of known wavelength is directed on to the crystal under investigation. The beam is split up by the crystal into many beams, each the result of diffraction by a plane of atoms in the crystal. For a particular X-ray beam, the angle of diffraction θ is related to an atomic spacing d within the crystal (Fig. 14). By recording the angles through which an X-ray beam is diffracted, it is possible to determine the atomic plane spacings within the mineral concerned and thus build up a picture of its atomic structure.

There is such a close relationship between atomic structure and chemical composition that it is possible indirectly to determine the major element composition of some minerals simply by making measurements of their atomic structure. An example of this is provided by the olivines. These vary in chemical composition between pure iron silicate (Fe_2SiO_4—fayalite) and pure magnesium silicate (Mg_2SiO_4—forsterite). All possible intermediate ratios of Mg to Fe are possible. Now, because the Fe^{2+} ion is larger (r=0.74 Å) than the Mg^{2+} ion (r=0.66 Å), the atomic structure of olivines rich in iron is slightly more distended or open than that of olivines rich in magnesium. Accordingly, atomic planes are more widely spaced in Fe-rich olivines than in Mg-rich olivines. It is thus possible, by direct measurement of the distance between suitable planes, to determine the chemical composition of an olivine sample.

SAQ 23 A beam of X-rays with a wavelength of 1.54 Å is directed on to a crystalline sample of an olivine of unknown composition. One of the X-ray beams is diffracted away from the sample at an angle θ of 15.9°. Use the Bragg equation to calculate the d spacing of the atomic planes responsible for this diffraction (take n=1). Figure 17 shows the relationships of the distance between these atomic planes to the chemical composition of the olivine. What is the chemical composition of an olivine with the d spacing you have determined?

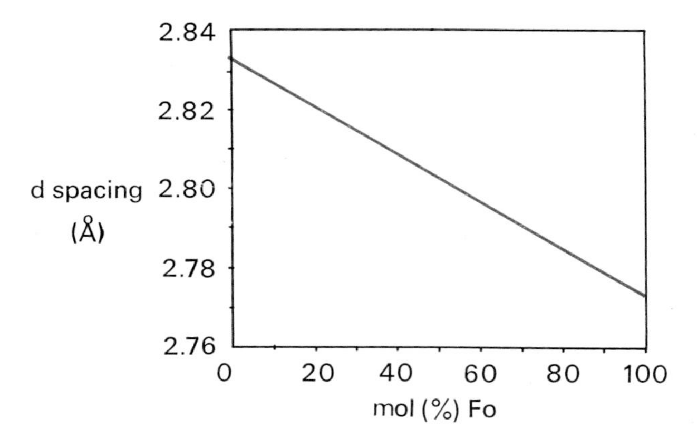

Figure 17 Relationship between d *spacing of olivines and their composition in terms of Mg_2SiO_4 (Fo) and Fe_2SiO_4 (Fa).*

Since it is fairly easy to do an X-ray analysis of olivine samples, this method is commonly used for the major element analysis of olivines. Remember, however, that since this method depends on a simple relationship between physical structure and chemical composition, it can only be used for simple mineral groups, of which the olivines are a good example.

SAQ 24 Which method of analysis would you use for the following problems? Indicate the possible choices by using the letters indicated.

1 Fluorine in mica.
2 The Sr^{87}/Sr^{86} ratio of an ultrabasic rock (peridotite).
3 All the major elements in a large number of samples from a particular area.
4 Thorium in a single meteorite, where it is usually present at concentrations of below 1 ppm.
5 Boron in a large number of soil samples.
6 Cerium in a basalt where it is usually present at below 100 ppm.
7 All the major elements in a set of standards for use with a physical method of analysis.

A Optical emission.
B X-ray fluorescence.
C Mass spectrometry.
D Wet chemical analysis.

1.8 Summary

Geochemistry is the study of the distribution of the chemical elements throughout the Earth. Since much geochemical data comes from samples of rock collected at the Earth's surface, it is necessary to ensure that the samples collected are representative of the rocks. A variety of methods are used to analyse collected

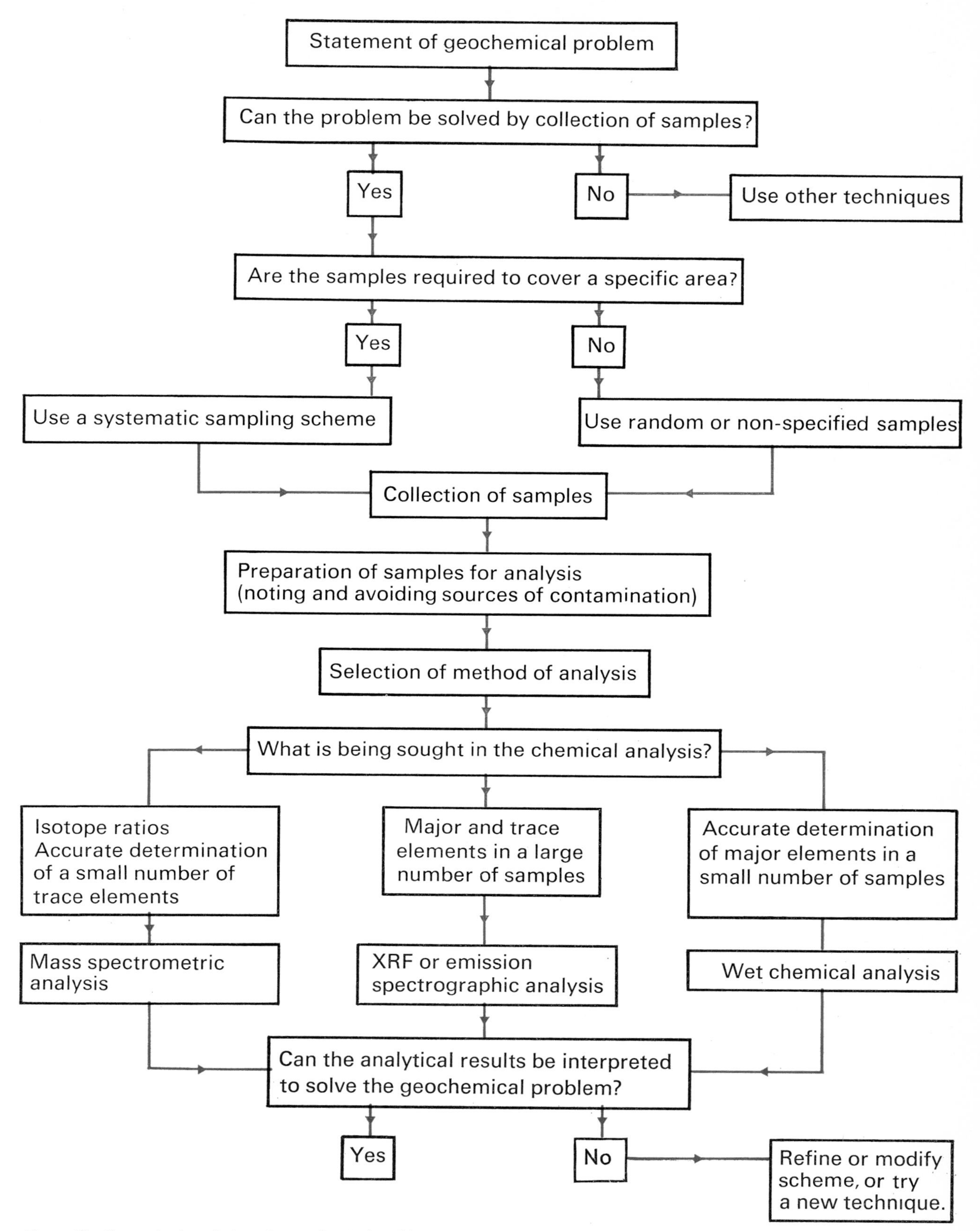

Figure 18 Stages in the solution of a geochemical problem.

samples, including wet chemical techniques and emission, X-ray and mass-spectrometric methods. Figure 18 provides a summary of the stages involved in a typical geochemical project.

Many attempts have been made to calculate the bulk composition of the Earth's crust. A review of these attempts suggests that the upper continental crust has a composition between that of basalt and granite, but closer to the latter when most elements are considered. This is an important conclusion since it helps us to evaluate different models for the whole continental crust (Unit 2), and its origin (Unit 5).

In this Unit we have been concerned with the collection of geochemical data. This descriptive aspect is important, as these results tell us about the other aspects of geochemistry—the processes and reactions which brought about the chemical compositions we measure. In the next Unit we complete our description of the geochemical features of the Earth and its minerals; we shall then be in a position to explore the processes of geochemical reactions.

Self-assessment Answers and Comments

SAQ 1 The correct answers are 2 and 5, although of course items 1, 3 and 4 are components of the discipline!

SAQ 2

(*a*) Because granite contains free SiO_2 in the form of the mineral quartz, which may make up as much as 30% of the rock. Basalt typically contains no quartz.
(*b*) Biotite and iron oxide for the granite; pyroxene, olivine and iron oxide for the basalt.
(*c*) Granite contains only about 5% of iron- and magnesium-bearing minerals, but they form almost half of any basalt. (See Table 1, p. 10.)
(*d*) SiO_2. All the minerals except iron oxide are silicate minerals.

SAQ 3

(*a*) The granite will contain

$$\frac{\overset{\text{(feldspar)}}{(66\times75)}+\overset{\text{(quartz)}}{(100\times25)}}{100}=74.5\%\ SiO_2$$

(*b*) We can be virtually certain that the analysis will fall between 74.56 and 74.44% SiO_2, i.e. between (74.5+0.06) and (74.5—0.06).

SAQ 4

(*a*) The value falls at a little above 1.75% TiO_2. We shall use 1.75% TiO_2 as a convenient figure. If you used 1.9% or thereabouts, your answer for (*b*) will be slightly different from ours.
(*b*) (10+31+154+84+49)=328 circumoceanic basalts out of the total of 360 have TiO_2 contents of *less than* 1.75% TiO_2. The point at 1.75 represents basalts with TiO_2 contents of 1.4–1.75%. That is about 11/12 or nearly 92%. Similar calculations show that some 94% of the oceanic basalts considered in this study have TiO_2 contents *over* 1.75%.

SAQ 5

(i) I 3
II 1, 2, 5
III 4

(ii) During the development of a soil from a parent rock, Ca, Mg, Na and K show a marked decrease, Fe and Al show a small decrease, while Si often shows an increase. Note, however, that Si may be leached under certain conditions, for example those causing the development of laterites in tropical climates.

SAQ 6 1 C; 2 A; 3 A; 4 B; 5 A; 6 D; 7 A.

SAQ 7

(*a*)

A	B	C
CaO, MgO, MnO	Al_2O_3*	SiO_2, Na_2O*
P_2O_5, SO_3	K_2O*	
organic matter		

* *From the point of view of significant differences, Al_2O_3, K_2O and Na_2O are enriched in both B and C relative to A. Fe_2O_3 and TiO_2 are not significantly enriched in any horizon of this profile.*

(*b*) From a quartz-rich sandstone. The analysis of horizon C is closest to that of the sandstone, column 1 of Table 6. See also the note of caution at end of Section 1.5.

SAQ 8 Parts per million expressed as percentages come out as follows:

O 46.6; Al 8.13; Si 27.7; Ti 0.44; Cr 0.01; Fe 5.0; Cu 0.0055; Pb 0.0013.

SAQ 9 The principal difficulties are in being certain that the sampling procedure is reliable, and that suitable weighting is given to different kinds of sample. This requires much geological information in the form of maps and cross-sections to provide data about relative volumes of different rock types. Only when data of this kind is truly reliable can we make valid assessments of whether there are significant compositional differences in different continental areas.

SAQ 10 (iii) is correct, according to the data in Table 6, and information from *Principles of Geochemistry*. As you will see, however, it is somewhat closer to granite than basalt, so (i is only slightly wrong!

SAQ 11 SiO_2.

SAQ 12 Oxygen.

SAQ 13 The ionic radius of oxygen (r=1.40Å) is much greater than that of silicon (r=0.42 Å). This means that the volume occupied by oxygen within the Earth's crust is much larger than silicon. This point will be expanded in Unit 2.

SAQ 14 We can assume the analyses were reliable! There may be differences in these components between the two regions, but it is unlikely because all the other components have such similar values. From what you have found out about soils, you might agree that 2 seems the most likely explanation.

SAQ 15

(*a*)

1		2		3		4	
O	46.6	O	85.9	N_2	75.5	O	62.8
Si	27.7	H	10.6	O_2	23.1	C	19.4
Al	8.1	Cl	1.9	Ar	1.3	H	9.3
Fe	5.0	Na	1.1	C*	0.01	N	5.1
Mg	2.1		——		——	Ca	1.4
Ca	3.6		99.5		99.9		——
Na	2.8						98.0
K	2.6						
	——						
	98.5						

* $wt.\%\ C\ in\ CO_2 = \frac{12}{44} \times 100 = 27\%$.

wt.% C in average composition of the atmosphere =

$$\frac{460 \times 10^2}{10^6} \times \frac{27}{10^2} = \frac{12 \times 10^3}{10^6} = 12 \times 10^{-3} \approx 0.01$$

(*b*) (i) 1:8; 2:3; 3:2; 4:5. A very good illustration of the importance of a few major elements! In fact, among all those columns, there are only *13* elements, out of the known 92 natural ones.

(ii) Oxygen.

(iii) Carbon of course is concentrated in the biosphere, for it is the foundation of all terrestrial living matter. Notice how much more dilute it is in the atmosphere—column 3 of (*a*).

SAQ 16

(*a*) Mean $\bar{x} = \frac{9+10+7+11}{4} = \frac{37}{4} = 9.25$ ppm

Standard deviation $s =$

$$\sqrt{\frac{(9-9.25)^2+(10-9.25)^2+(7-9.25)^2+(11-9.25)^2}{4}}$$

$$= \sqrt{\frac{0.25^2+0.75^2+2.25^2+1.75^2}{4}}$$

$$= \sqrt{\frac{8.75}{4}} = \sqrt{2.19} = 1.48 \text{ ppm}$$

The relative standard deviation or coefficient of variation =

$$C_V = \frac{100 \times 1.48}{9.25} = 15\%.$$

(*b*) We can now specify how closely figures taken by this method will approximate the mean. In this case, about 67% of the readings taken will have values of 9.25 ± 1.48 (s) ppm, and 95% will have values of 9.25 ± 2.96 ($2s$) ppm (and so on).

(*c*) In view of the statement that C_V for trace element analyses can be as low as 0.5%, the precision of these analyses ($C_V = 15\%$) is not particularly good; but it may well have been adequate for the particular problem involved.

SAQ 17

1 Since coverage of the whole of Cornwall is required, a *systematic* sampling plan would be used. However, to detect the areas *richest* in tin, only precise measurements would be needed.

2 Since we are not concerned with a particular area, it is better to collect samples as widely and *randomly* as possible. Again, only *precise* measurements are needed since we are not seeking absolute values.

3 *Systematic* sampling of the particular area. Since we need an average composition, *accurate* analysis would be necessary.

4 *Random* samples of mica from a great variety of rock types would be used, and since we are trying to establish averages, the analyses would need to be as *accurate* as possible.

SAQ 18 Calcium oxide $= \frac{20.8}{200} \times 100 = 10.4\%$ CaO.

Basalts typically have this order of CaO content.

SAQ 19 The line intensities, and hence the concentrations, for both K and Rb decrease in the order mica–granite–dolerite. Accurate values for K and Rb in the granite (G–1) and dolerite (W–1) are given in Table 3.3 (pp. 45–6) in *Principles of Geochemistry*.

SAQ 20 $I\ Cr_{4254}/I\ Pd_{4473} = \frac{12}{600} = 0.02$

From Figure 11, Cr = 8 ppm. This is closer to the average content of Cr in dolomitic limestones than in ultrabasic igneous rocks. An origin from the former is more likely.

SAQ 21 $n\lambda = 2d \sin\theta$

$7.13 = 8.74 \sin\theta$

$$\sin\theta = \frac{7.13}{8.74} = 0.8158$$

(Hence $\theta = 54°40'$ (or 54.67°).

SAQ 22

(*a*) The count ratios are:

Basalt I 1992/996 = 2.0
Basalt II 4040/1010 = 4.0

(*b*) Using the equation given, and the graph, Figure 16:

Basalt I $TiO_2 = 1.33\%$
Basalt II $TiO_2 = 2.87\%$

(*c*) Subject to limitations discussed in Section 1.4.2, basalt I is likely to be circumoceanic, whereas basalt II is likely to be of oceanic type.

(*d*) For the first basalt, the *spread* is between about 1.65 and 1.0. The TiO_2 content from (*b*) is 1.33%, so $3s \approx 0.33$, so s is 0.11%. For the second basalt, the *spread* is between about 3.25 and 2.45. The TiO_2 content from (*b*) is 2.87%, so $3s \approx 0.4$, so s is 0.13%. These are not enough to affect the classification of basalt I as circumoceanic and basalt II as oceanic.

SAQ 23 $n\lambda = 2d \sin\theta$

$1.54 = 2d \sin 15.9°$

$$d = \frac{1.54}{0.2740 \times 2} = 2.81\text{Å}$$

From Figure 18, such an olivine contains 30% forsterite.

SAQ 24 1A, 2C, 3B, 4C, 5A, 6B, 7D. If you had any difficulty look back at the appropriate Section in the Main Text and Table 9.

Appendix 1

References to S100*

1 Unit 24, *Major Features of the Earth's Surface*, Section 2 (including Plate A).
2 Unit 14, *The Chemistry and Structure of the Cell*, Section 2.
3 Unit 20, *Species and Populations*, Section 3.2.
4 Unit 22, *The Earth: Its Shape, Internal Structure and Composition*, Section 5.8.
5 Unit 10, *Covalent Compounds*, Section 3.
6 Unit E, *Handling of Experimental Data*, Sections 1 and 2.1–2.6.
7 Unit 6, *Atoms, Elements and Isotopes: Atomic Structure.*
8 TV24.
9 Unit 11, *Chemical Reactions*, Home Experiment.
10 Unit 7, *Electronic Structure of Atoms*, Section 1.2.
11 Unit 6, TV 6.
12 Unit 6, Section 2.5.

Appendix 2

Glossary

ALKALI FELDSPAR A feldspar typical of granites and rich in the alkali elements Na and K. Chemical compositions vary between albite ($NaAlSi_3O_8$) and orthoclase ($KAlSi_3O_8$). See also Table 2 in Main Text.

ALKALINE BASALT A variety of basalt, abundant on oceanic islands, which is distinguished from subalkaline basalts (*q.v.*) by a lower content of SiO_2 and a higher TiO_2 and alkali (K_2O+Na_2O) content.

ANDESITE A volcanic igneous rock of intermediate chemical composition, containing about 60% SiO_2, typical of island arc volcanism.

AUGITE A pyroxene (*q.v.*) very common in basic rocks (basalt), also found in some intermediate rocks (diorite, *q.v.*) and having the formula $Ca(Mg, Fe)Si_2O_6$.

BIOTITE A mica (*q.v.*) common in granites and schists. Biotite is a dark mica due to its content of Mg and Fe, and is also a hydrous mineral. Note the (OH) groups in the formula $K(MgFe)_3(SiAl)_4O_{10}(OH)_2$.

CANADIAN SHIELD A large area of ancient rocks, mainly igneous and metamorphic, which occur at or near the surface throughout much of Canada.

CLAY MINERALS Hydrated aluminosilicates which occur abundantly as minute platy particles in sedimentary rocks, particularly shales. The chemical compositions are rich in alumina and water, as is seen from the formula of one example—kaolinite $Al_4Si_4O_{10}(OH)_8$. More information about clay minerals is presented in Unit 2.

DACITE A volcanic rock with the same chemical composition as granodiorite (*q.v.*).

DIORITE An intrusive igneous rock of intermediate chemical composition and chemically equivalent to the lava andesite (*q.v.*).

* *The Open University (1971) S100* Science: A Foundation Course, *The Open University Press.*

DOLERITE An intrusive igneous rock similar to basalt in composition but slightly coarser-grained.

DOLOMITIC LIMESTONE Limestone with a high proportion of dolomite—a mixed carbonate which contains magnesium and has the formula $CaMg(CO_3)_2$.

GARNET A dense silicate which is characteristic of metamorphic rocks formed under high pressures and temperatures. A common type in such rocks is almandine, $Fe_3Al_2Si_3O_{12}$. Garnets may be used as an abrasive or as a gemstone.

GRANODIORITE An intrusive igneous rock, similar to diorite (*q.v.*) but containing more quartz. This gives a slightly higher SiO_2 content, and a chemical analysis of a granodiorite may contain about 66% SiO_2. It thus lies on the borderline between intermediate and acid compositions.

HORNBLENDE An amphibole common in intermediate rocks (diorites and granodiorites). The chemical composition is similar to that of pyroxenes but amphiboles are hydrous. The simplified formula of a common hornblende is $Ca_2(Mg, Fe)_5 Si_8O_{22}(OH)_2$.

IRON OXIDE Almost all rocks contain small amounts of iron oxide. Magnetite (Fe_3O_4) and ilmenite ($FeTiO_3$) are common in igneous and metamorphic rocks; haematite (Fe_2O_3) is common in soils and sedimentary rocks, often in hydrated form as limonite, $Fe_2O_3.nH_2O$, where n is generally not greater than 3.

MICA A hydrated aluminosilicate common in sedimentary and acid igneous rocks, where it occurs in typical flake shaped grains. There are two chief varieties —biotite (*q.v.*), the dark mica, and muscovite, which is white in colour because it contains no MgFe, its formula being $KAl_2(SiAl)_4O_{10}(OH)_2$ (*cf.* biotite).

MINERAL Naturally occurring inorganic compound possessing a definite crystalline structure and chemical composition varying between fixed limits. Physical aggregates of minerals are rocks.

OLIVINE An Mg–Fe silicate common in basic and ultrabasic rocks. The chemical composition varies between forsterite (Mg_2SiO_4) and fayalite (Fe_2SiO_4).

PLAGIOCLASE FELDSPAR A feldspar containing Na and Ca. Chemical compositions vary between albite ($NaAlSi_3O_8$) and anorthite ($CaAl_2Si_2O_8$), end-members of a solid solution series which will be described further in Unit 2.

PYROXENE An Mg–Fe silicate common in basic igneous rocks. Chemical compositions may vary between $MgSiO_3$ and $FeSiO_3$ or may contain calcium, when the pyroxene would have the composition of augite (*q.v.*).

SERPENTINITE A rock composed essentially of serpentine, a hydrated magnesium silicate, $Mg_3(Si_2O_5)(OH)_4$. Serpentine is often derived by the hydration of magnesium olivine (*q.v.*). Thus serpentinites can be derived from the alteration and hydration of peridotite. It can also be derived by alteration of other magnesium-rich rocks such as dolomitic limestones (*q.v.*).

SPECTROPHOTOMETER An analytical instrument similar in principle to a colorimeter but designed to measure the intensity of light of fixed wavelength. You may have used one during your S100 Summer School.

SUBALKALINE BASALT (=tholeiitic basalt) The most common basalt type occurring on the ocean floor and on the continents, distinctive for its low content of alkalis (Na_2O+K_2O) and TiO_2.

WEIGHTED AVERAGE Average calculated from a large number of data by using each piece of data in proportion to its overall importance.

Appendix 3

The Bragg equation $n\lambda = 2d \sin \theta$

The derivation of this equation is given below. To follow this you will need to refer back to Figure 14, looking carefully at its caption.

This Figure shows a parallel beam of X-rays (of wavelength $= \lambda$) impinging on a crystal which consists of atoms arranged in parallel planes a distance d apart. The beam is as wide as d (i.e. $Y'Y = d$). You can see that the ray XYZ travels a distance CYD greater than the ray X′Y′Z′.

Now, for the two rays to reinforce each other and be diffracted, the distance CYD must be equal to a whole number of wavelengths of the incident beam. If this condition is not fulfilled, the two rays interfere with each other and the resultant intensity is reduced.

To determine the conditions for diffraction we need to find CYD.

$$\sin \theta = \frac{YD}{Y'Y}$$

$$\text{but } Y'Y = d$$

$$\text{therefore } \sin \theta = \frac{YD}{d} \text{ and } YD = CY = d \sin \theta$$

$$\text{therefore the distance } CYD = 2d \sin \theta$$

Diffraction will occur when $n\lambda = 2d \sin \theta$, where n is an integer. When $n = 1$, the reflected beam is called a first-order diffraction. Similarly, $n = 2$, the diffraction will be second-order, and so on.

When a beam of X-rays of mixed wavelength is directed on to a crystal, only those wavelengths which fulfil the Bragg law will be reflected.

Acknowledgements

Grateful acknowledgement is made to the following for material used in this Unit:

Figures 1 and 2: *Geologisia Foreningen*, **508,** 1962; Figure 3: The Mineralogical Society for F. Chayes, 'Titania and alumina content of oceanic and circum-oceanic basalt' in *Mineralogical Magazine*, **34,** 1965; Figures 4a and 4b: US Department of Agriculture, Soil Conservation Service; Figures 6a and 6b: National Research Council of Canada for D. M. Shaw *et al.* 'An estimate of the chemical composition of the Canadian Precambian Shield' in *Canadian Journal of Earth Sciences*, **4,** 1967; Figure 7: W. H. Freeman & Co for Howell Williams *et al. Petrography*, Copyright © 1954; Figure 9b: Rank Precision Instruments and Hilger & Watts; Figure 10: Interscience Publishers Inc. for S. R. Taylor and L. H. Ahrens, 'Spectrochemical analysis' in *Methods of Geochemistry*, A. A. Smales and L. R. Wager (ed.), Copyright © 1960; Figures 15 and 16: Elsevier Publishing Co for B. E. Leake *et al.*, *The Chemical Analysis of Rock Powders by Automatic X-ray Fluorescence;* Figure 17: Mineralogical Society for J. V. Smith and R. C. Stenstrom, 'Chemical analysis of olivines by the electron microprobe' in *Mineralogical Magazine*, **34,** 1965.

UNIT 2 COMPOSITION AND STRUCTURE OF THE EARTH AND ITS MINERALS

Contents

Table A

List of Scientific Terms, Concepts and Principles used in Unit 2

Taken as prerequisites				**Introduced in this Unit**			
Assumed from general knowledge (GK) or from S100*	**Unit No.**	**Defined in a previous Unit in S2–2**	**Unit No.**	**Defined and developed in this Unit**	**Page No.**	**Introduced here, but fully developed in later Unit**	**Unit No.**
acid	9	collection of rock samples	1	silicate tetrahedron	7	iron meteorites	3
anion	8	granodiorite model for upper crust	1	silicate anion	8	sulphur in the Earth's core	3
asbestos	GK	acid rock	1	aluminate anion	8	partial melting	4
base	9	intermediate rock	1	polyanions	8	fractional crystallization (or crystal fractionation)	4
clay	GK	basic rock	1	radius ratio	9		
constructive plate margin	25	ultrabasic rock	1	coordination number	9		
core	22			coordination polyhedra	10		
Conrad discontinuity	22			cubic close-packing	12		
covalent bond	8			hexagonal close-packing	12		
crystal lattice	5			isomorphism	13		
destructive plate margin	25			ionic substitution	13		
diamond structure	8			solid solution	13		
earthquake focus	22			coupled substitution	13		
electronegativity	8			polymorphism	14		
graphite structure	8			silicate polytetrahedra	14		
ionic array	8			oxygen sharing	14		
ionic (electrovalent) bond	8			orthosilicate	16		
ionic radius	8			olivine structure	16		
island arc	24			ring silicate	17		
mantle	22			chain silicate	18		
metamorphism	24			cleavage	18		
noble gas	8			band silicate	18		
oceanic crust	22			layer silicate	19		
oceanic ridge	24			talc structure	20		
P-wave velocity	22			mica structure	21		
peridotite	22			framework silicate	22		
plate tectonics	25			silica structure	22		
salt	9			feldspar structure	23		
seismic wave velocity	22			alkali feldspars	23		
				plagioclase feldspars	24		

* *The Open University (1971)* S100 Science: A Foundation Course, *The Open University Press.*

Taken as prerequisites				Introduced in this Unit			
Assumed from general knowledge (GK) or from S100*	**Unit No.**	**Defined in a previous Unit in S2–2**	**Unit No.**	**Defined and developed in this Unit**	**Page No.**	**Introduced here, but fully developed in later Unit**	**Unit No.**
sial	22			clay minerals	26		
sima	22			kaolinite	27		
tetrahedron	GK			montmorillonite	27		
X-ray diffraction	28			non-silicate minerals	28		
				oceanic tholeiite	30		
				ocean crust structure	31		
				pillow lavas	31		
				amphibolite	33		
				eclogite	33		
				granulite model for lower crust	34		
				chemical composition of granulite	34		
				heat producing elements	35		
				ultramafic nodules	40		
				olivine nodules	40		
				garnet peridotite	40		
				kimberlite	40		
				Alpine-type peridotite	41		
				depleted and undepleted mantle	43		
				pyrolite model	44		
				mineral reactions in the mantle	44		
				Fe/Ni/light element model for core	46		

Conceptual Diagram

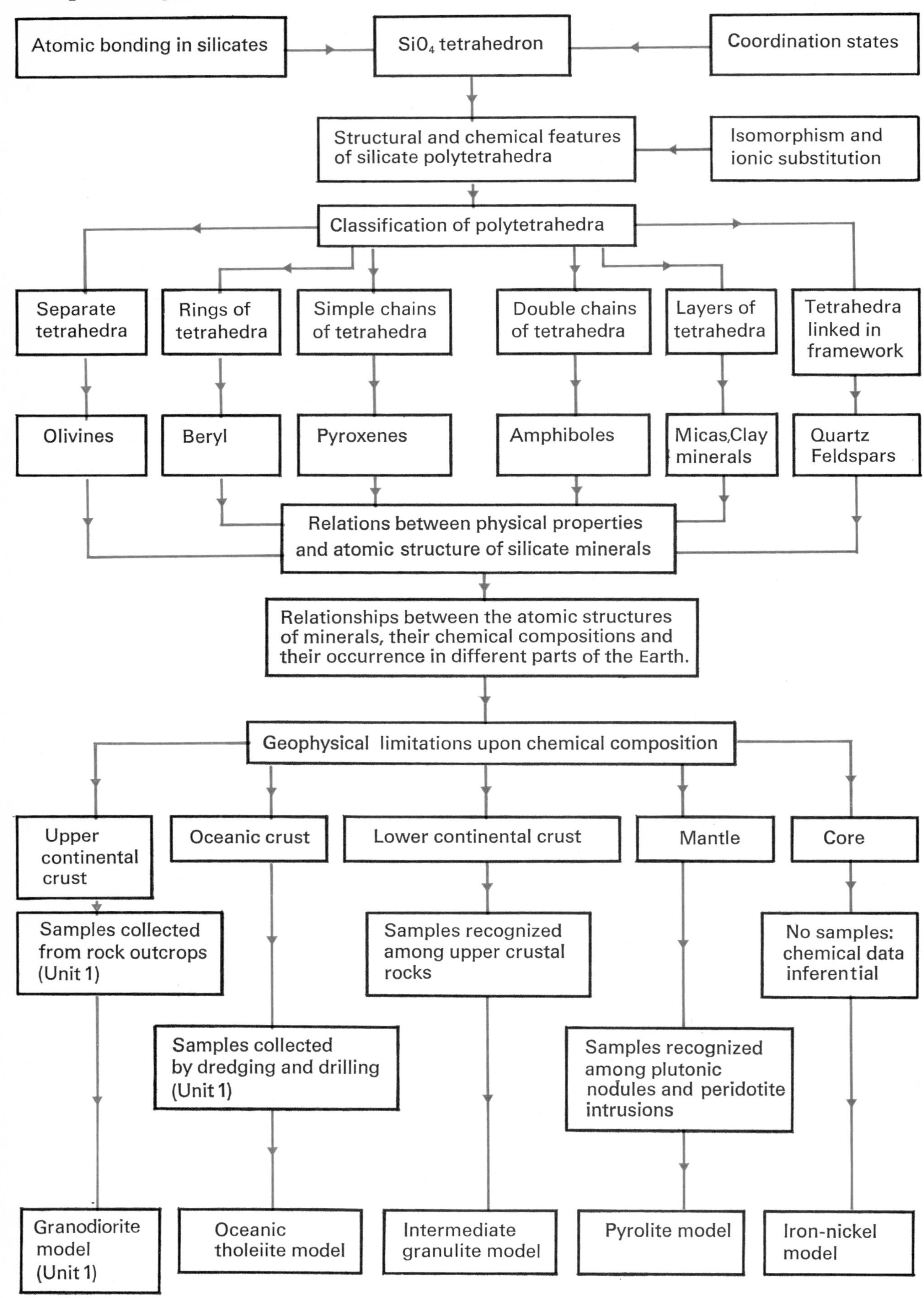

Objectives

When you have completed this Unit you should be able to:

1 Define, or recognize definitions of, or distinguish between true and false statements concerning each of the terms, concepts and principles in the third column of Table A.

2 Describe the features and the reasons for the importance of the SiO_4 tetrahedron.

3 Explain the relationship between ionic radius and coordination number in crystal structure.

4 Distinguish between isomorphism and polymorphism and recognize the factors which influence ionic substitution in minerals.

5 Build simple models of silicate polytetrahedra with the Home Experiment Kit, and describe the structures of silicate polyanions as SiO_4 tetrahedra linked by oxygen sharing.

6 Explain the distinguishing structural features of the seven main groups of silicates.

7 Suggest how the physical properties of silicate minerals may be related to their molecular structure.

8 Explain which of the principles relevant to the study of silicate structures can also be applied to non-silicate structures.

9 Describe the overall physical features of the oceanic crust and label a model showing the relationships between its constituent parts. State how the chemical composition of ocean-floor basalt differs from that of continental basalts.

10 Explain why there are alternative models for the composition of the lower continental crust.

11 Given the granulite model for the lower continental crust, state how its chemical composition may differ from that of the upper continental crust. Explain one process which may account for the chemical difference.

12 List evidence favouring a peridotite model for the upper mantle.

13 Given chemical data on possible samples from the upper mantle, namely olivine nodules, Alpine-type peridotites and garnet peridotite nodules, show how this data can be used to categorize these samples as depleted or undepleted samples of a peridotitic upper mantle.

14 Outline the pyrolite model for the mantle and explain the changes that may take place with increasing depth in such a mantle.

15 Outline, giving reasons, the main chemical features of the Earth's core.

2.1 Introduction to Silicate Minerals *(Objectives 1, 2)*

In the first Unit we looked at the problems and methods of geochemical analysis, using as examples the most easily accessible parts of the Earth, especially the upper continental crust.

The principal constituents of the rocks in this outer part of the Earth are the silicate minerals. Since they also dominate the rest of the crust and the mantle, it is appropriate that we start this Unit by studying them in more detail, as well as looking at some important non-silicate minerals. We shall then be in a better position to examine the nature of the crust as a whole, as well as that of the mantle.

We shall consider how the pattern that runs through the structures of the different silicate minerals determines their physical and chemical properties. We shall see, indeed, how the occurrence of the silicates is related to the structural pattern: some structures are more open and less dense than others, and are more likely to be found in the crust than in the mantle. We shall also consider how their structure determines the way in which minerals are altered by natural processes, such as weathering, and how these processes, in their turn, determine the structures of the products.

The silicates form a flexible system encompassing a wide and rich variety, from fibrous asbestos to sheet-like mica, from soft talc (soapstone, or talcum powder) to hard quartz, from deep-blue lapis lazuli (ultramarine) to deep-red garnet, and from sand and clay to granite. This is not to mention the man-made materials, the glasses, pottery and porcelain, bricks and mortar and concrete.

The variety of chemical composition among silicates was quite bewildering to the older geochemists dealing with minerals having formulae like $(Mg, Fe)_2SiO_4$, $(K, Na) AlSi_3O_8$, and $K(MgFe)_3AlSi_3O_{10}(OH)_2$. They tried to apply the Law of Constant Proportions†, which had been so successful in the study of molecular and ionic compounds, such as acids, bases, and salts. They tried to classify the silicates as salts of hypothetical silicic acids, such as H_4SiO_4(ortho), H_2SiO_3 (meta), and very many polysilicic acids, starting with $H_2Si_2O_5$. But this was a thankless task, for the various acids could not be characterized, and many of the 'salts' did not fit the classification. In addition, compounds that should, from their compositions, have been similar, were not; while others that seemed to be chemically unrelated, showed puzzling similarities in crystal shape.

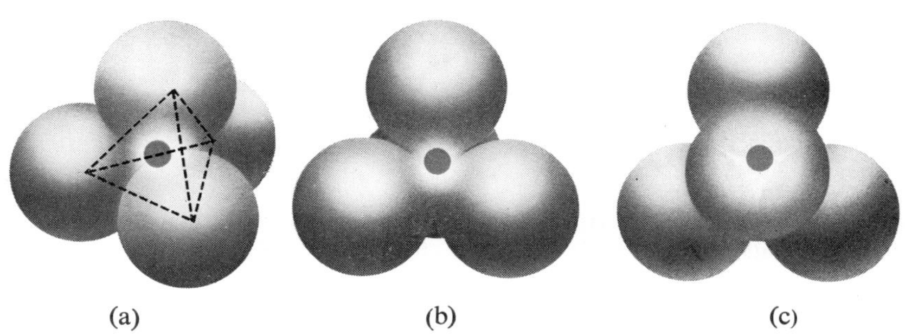

Figure 1 Three views of an SiO_4 tetrahedron. The large spheres are oxygen and the small red sphere in the centre is silicon.

The problem was intractable in those terms. But when X-ray diffraction crystallography was applied to silicate structures, notably by W. L. Bragg and by L. Pauling in the 1920s, the solution was remarkably simple. We shall explore it in depth in this Unit, and shall summarize it briefly here: The basis of all silicate structures was found to be a simple tetrahedron, SiO_4 (Fig. 1). Tetrahedra can share vertices to form different repeating patterns. The ions in the crystal lattices are readily substituted by others of the same size, with perhaps a different charge;

silicate tetrahedron

† *Items symbolized thus are defined in the Glossary, Appendix 2.*

the charges will then be balanced by other substitutions. We shall see how, with the help of a few simple rules, the remarkable variety of form and composition of the silicates can be understood.

2.2 Bonding in the Silicates – Covalent and Ionic *(Objectives 1, 2)*

As you learned in the Science Foundation Course (S100[1])* electronegativities of elements can be used to predict what kinds of bonds they will form with one another. The electronegativity of oxygen is given as 3.5, and that of silicon as 1.8. The difference between the two is such that the Si–O bond is about 50–50 covalent-ionic. We can think of it either as a covalent bond with ionic character or the other way about. (We should note here that we are modifying a statement made in S100[2] about the nature of the Si–O bond: we said then that it was entirely covalent, which was an oversimplification.) Mineralogists and crystallographers describe silicates as assemblages of Si^{4+} and O^{2-} and other ions, but this is an exaggeration. For silicon, the ionization energy required to remove four electrons from the atom is quite unattainable (cf. carbon, S100[3]), and this gives covalent character to the bonds.

It is most convenient, however, to consider all the elements in silicate structures to be present in their ionic configuration (i.e. that of the nearest noble gas, S100[4]).

SAQ 1 Refer to the Table provided in S100[5], and state which noble gas has the same electronic configuration as the ions Si^{4+}, Al^{3+} and O^{2-}.

Although the Si–O bond has a large covalent component, the highly stable SiO_4 tetrahedral arrangement (Fig. 1) is in fact a complex anion:

$$Si^{4+} + 4O^{2-} \rightarrow \underset{\textit{silicate anion}}{(SiO_4)^{4-}}$$

silicate anion

Aluminium is *amphoteric*; that is, it can form both acidic and basic oxides and can therefore have 'non-metallic' as well as metallic properties.

It can thus have two different roles in crystal structures, as the cation Al^{3+}, and also as part of another complex anion:

$$Al^{3+} + 4O^{2-} \rightarrow \underset{\textit{aluminate anion}}{(AlO_4)^{5-}}$$

aluminate anion

Indeed, both the cation and the anion can be found in the same structure, for example muscovite mica, where the aluminate ion substitutes, in part, for silicate. We can compare these complex anions with phosphate, sulphate, and chlorate, the later members of the isoelectronic series† which runs from $AlO_4{}^{5-}$ to $SiO_4{}^{4-}$, $PO_4{}^{3-}$, $SO_4{}^{2-}$ and $ClO_4{}^{-}$. They all have strong covalent internal bonds, a tetrahedral arrangement as in Figure 1, and persist unchanged through many reactions.

There is, however, one important respect in which aluminate and silicate differ from the others – in their ability to link up by oxygen-sharing, to form very stable *polyanions* of varying size. Irrespective of whether these polyanions are small or large, they are held together in the structure by metal cations. Together they form an *ionic array*, as do salts such as sodium chloride (S100[6]). For the purposes of examining the geometry of silicate structures in detail, however, it is more useful to consider all the elements present as ions whose size and charge determine their spatial relationships.

polyanions

* *The Open University (1971)* S100 Science: A Foundation Course. *The Open University Press. Superscript numbers indicate more detailed references to S100, which are listed in Appendix 1.*

2.3 Ionic Radius and Coordination Number (*Objectives 1, 3, 5*)

You should now read the section, 'Principles of Crystal Structure', in Principles of Geochemistry* *(p. 75 to the top of p. 81). Then return to this text, which now elaborates the more important points.*

1 As we have indicated in Unit 1, the large size and abundance of the oxygen ion means that silicate structures are mainly a packing of oxygen ions with cations in the interstices—which is why the lithosphere is also called the oxysphere.

2 Table 1 provides a pictorial supplement for the more important ions, whose ionic radii are presented in Figure 4.1 of *Principles of Geochemistry*. It illustrates well the disparity in size between oxygen and silicon, the two most important elements in the crust.

Table 1 Some ionic radii in ångströms*

Li^{+} 0·68	Be^{2+} 0·35		B^{3+} 0·23		O^{2-} 1·40	F^{-} 1·36
Na^{+} 0·97	Mg^{2+} 0·66		Al^{3+} 0·51	Si^{4+} 0·42	OH^{-} 1·40	Cl^{-} 1·81
K^{+} 1·33	Ca^{2+} 0·99	Fe^{2+} 0·74 Zn^{2+} 0·74				Br^{-} 1·95
Rb^{+} 1·47	Sr^{2+} 1·12					I^{-} 2·16

* These figures refer, strictly speaking, to ionic bonding and to 6-fold coordination (see Appendix 3 (Black)). (Data from *Principles of Geochemistry*, Appendix I and Figure 4.1.)

SAQ 2

(a) What is the ratio of the radii of Si^{4+} and O^{2-}?

(*b*) What is the ratio of the volumes of Si^{4+} and O^{2-}? (Volume $= \frac{4}{3}\pi r^3$).

radius ratio

coordination number

This question helps to introduce the concept of *radius ratio*, which controls the *coordination number* and is the most important factor in determining the geometry of silicate structures.

3 The second most important factor in determining silicate structures is, of course, the ionic charge (Note: In *Principles of Geochemistry*, valency is virtually the same as electrovalency as defined in S100[7]).

Figure 4.2 in *Principles of Geochemistry* shows the principle of coordination

* *Brian Mason (1966)* The Principles of Geochemistry, *3rd ed. John Wiley (paperback).*

numbers in two dimensions. Our Figure 2 shows the *coordination polyhedra* to be found in silicate minerals (actually threefold coordination is not found in silicates, but typifies the carbonates with the CO_3^{2-} anion).

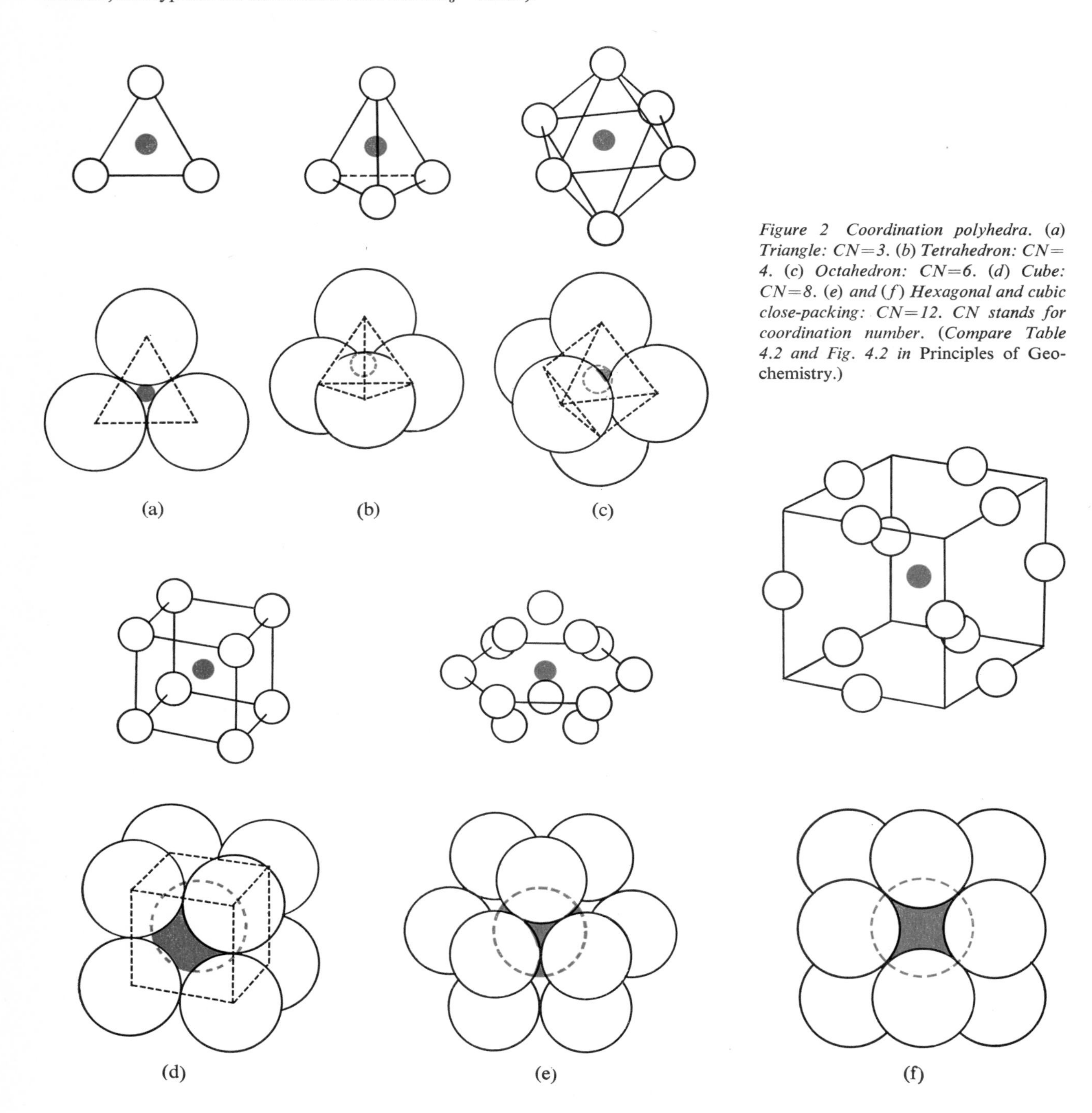

Figure 2 Coordination polyhedra. (a) Triangle: CN=3. (b) Tetrahedron: CN=4. (c) Octahedron: CN=6. (d) Cube: CN=8. (e) and (f) Hexagonal and cubic close-packing: CN=12. CN stands for coordination number. (Compare Table 4.2 and Fig. 4.2 in Principles of Geochemistry.)

Model Exercise A

You are provided with plastic balls, steel spokes and plasticine. For this experiment you will need 8 plastic balls (any colour) and the plasticine. Take the plasticine and, by rolling it between your palms, form four balls with diameters as near to 3.0, 4.5, 8.5 and 16.0 mm as you can. Now, for each of these plasticine balls, try to find out the *maximum number* of plastic balls which can be fitted together so that these balls *just touch* the plasticine ball. You can get some help from Figure 2, but your main problem will be to get three-dimensional arrays of balls to stay together! Enter your results in the two blank columns of the following Table.

Size of plasticine ball (mm)	*Calculate, and insert the ratio of diameters*, plasticine ball to plastic ball*	*Maximum number of plastic balls just touching the plasticine ball*
3.0		
4.5		
8.5		
16.0		

* Note that the plastic balls have a diameter of approximately 21 mm.

The numbers you should have entered in the last column are the coordination numbers which correspond to the radius ratios you have calculated in the second column. Compare them with Table 4.2 in *Principles of Geochemistry.*

The tetrahedron, octahedron, and cube, with the dodecahedron and icosahedron, are the five Platonic solids. The Pythagoreans, who considered numbers and geometry to be the basis of the Universe, took these polyhedra to be the five elements, respectively fire, air, earth, ether (spirit), and water. These ideas were expounded by Plato as examples of the beauty and significance of number and perfect form. The Platonic solids have again come into their own in the microcosm; that is, in molecular and crystal structure. A coordination polyhedron (as opposed to spherical ions) should, by repetition, fill space without leaving any gaps. This is true of the octahedron and cube, though not of the tetrahedron by itself. But tetrahedra and octahedra packed together will fill space, as they do in many silicate structures.

What about other simple forms of coordination? Could 5– or 7–coordination polyhedra fill space without leaving gaps? You will see that this is unlikely if you try fitting regular two-dimensional pentagons and heptagons together. (If you are interested in the actual geometrical calculations involved in working out coordination from radius ratios, we suggest you now turn to Appendix 3, treating it as black-page material.)

This review of radius ratios enables us to understand, on the one hand, how aluminium is able to form aluminate anions analogous to silicate anions and, on the other, how it can also occur as a 6-coordinated cation, replacing other cations such as Fe^{2+} or Mg^{2+}. This behaviour is important because aluminium is the most abundant element after oxygen and silicon.

SAQ 3 Why can aluminium occur in both 4- and 6-coordination with oxygen?

When alternative coordination numbers are possible for an ion, the one adopted may be influenced by the temperature and pressure at which the mineral crystallizes.

What physical conditions favour a high coordination number?

Lower temperatures and/or higher pressures favour higher coordination numbers since these give structures of smaller volume. Thus in minerals formed at lower temperatures and/or higher pressures, aluminium tends to 6-coordination, but at higher temperatures and/or lower pressures it tends to 4-coordination, replacing silicon to form more open aluminosilicate structures (p. 78, *Principles of Geochemistry*).

In mineral formation, the effect of higher pressure to increase coordination numbers is more important than the effect of temperature. It is thought that deeper in the mantle, silicon becomes 6-coordinate in structures of the spinel*

* *Spinel is $MgAl_2O_4$, in which Mg^{2+} is tetrahedrally coordinated and Al^{3+} is octahedrally coordinated by oxygen, each oxygen being bound to one Mg^{2+} and three Al^{3+}. In high-pressure silicate spinel structures, the Si^{4+} occurs in the octahedral sites, which are occupied by Al^{3+} in the normal spinel structure.*

type and possibly in a very dense modification of quartz called stishovite (see p. 79 in *Principles of Geochemistry* and Section 2.14.4).

2.3.1 Close-packing of identical spheres

In Figure 2, coordination polyhedra for the closest packing of identical spheres have been included for comparison because, as you will see, close-packing of oxygens is found in some silicate structures. The negatively charged oxygens are held together by the cations, including Si^{4+}, in the holes or interstices between them. Figure 3 demonstrates in more detail the close-packing of spheres.

Does the arrangement shown in Figure 3a look familiar?

It is the same as that of *either* Na^{+} *or* Cl^{-} in the NaCl structure (S100[6]). In Figure 3b some of the spheres have been removed to show the close-packed layers perpendicular to a cube diagonal. The layers parallel to the faces of the cube are not close-packed.

cubic close-packing

Figure 3c shows the other form of close-packing, which has hexagonal symmetry, because the close-packed layers are now horizontal. Figure 3d shows a part of a single close-packed payer.

hexagonal close-packing

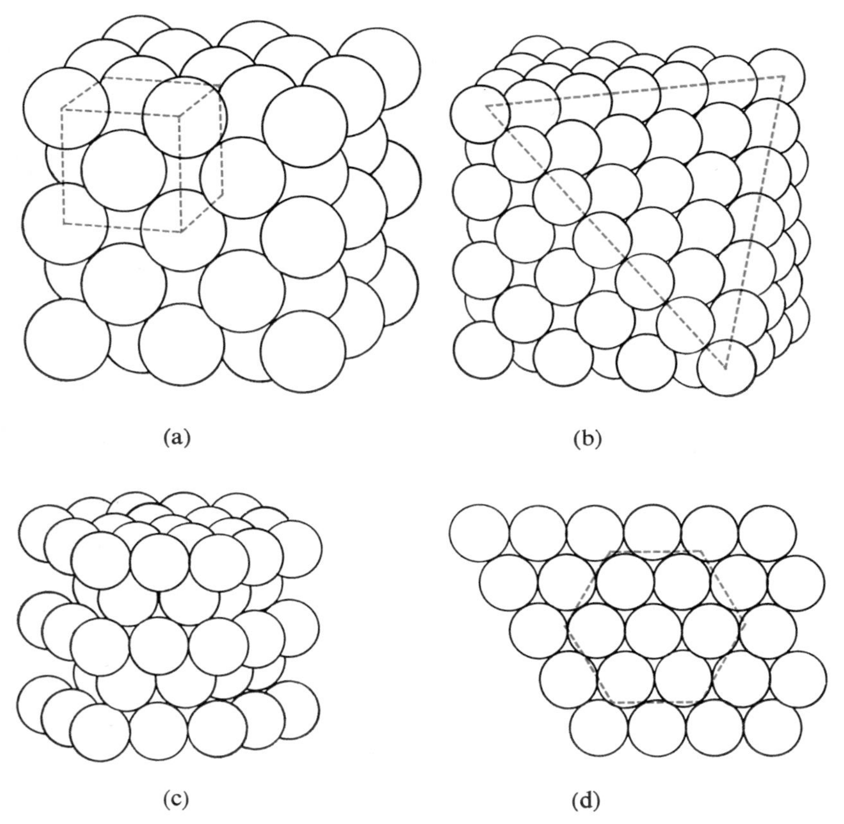

Figure 3 The close-packing of identical spheres. (*See text for discussion.*)

If you wish, build a close-packed structure using the plastic balls and the tray from your Home Kit. You can obtain cubic close-packing (as in Figs. 3a and b) if you put the third, fifth, seventh, . . . layers vertically above the first; i.e.

$$\begin{matrix} & 5 \\ 4 & \\ & 3 \\ 2 & \\ & 1 \end{matrix}$$

The distinction between the two forms of close-packing is less important, for our present purposes, than the next exercise, which is to examine the holes between the spheres. If you take a triangle of spheres in a close-packed layer, and put another sphere on it, you will have made a *tetrahedrally coordinated hole*. Can you see that this is not the only kind of hole in close-packed structures?

If you look, for example, at a hole in the side of the face-centred cube in Figure 3a and consider the grouping, you will see that it is part of an *octahedral cluster*, which would have one sphere in front of the hole and another behind. These are the two kinds of holes in the closest packing of spheres. In Section 2.7.1 we shall see that the olivine structure can be described very neatly in these terms. You can see this three-dimensionally if you make a two-dimensional *square* of balls, and then put another ball top and bottom, to look like the octahedron in Figure 2.

2.4 Isomorphism and Ionic Substitution *(Objectives 1, 4)*

You should now read from half way down p. 86 to half way down p. 90 in *Principles of Geochemistry* (Isomorphism and atomic substitution).

The most important points to note in this extract are:

1 The concept of *isomorphism* was useful to the early chemists in their attempts to determine valencies and atomic weights from equivalent (combining) weights. But their hypothesis that isomorphous salts have similar formulae (and vice versa) has many exceptions, and isomorphism itself is often difficult to define. With our present-day knowledge of crystal structure, older terms such as 'isomorphous', and 'solid solution', are being superseded by more precise terms such as 'isostructural', and 'ionic replacement'.

isomorphism

ionic substitution

solid solution

2 Ionic size is more important than ionic charge in substitution behaviour, because charge imbalances can always be made up by other substitutions elsewhere in the structure (*coupled substitution*). For example, if Si^{4+} is replaced by Al^{3+}, then Na^{+} in the crystal can be 'replaced' by Ca^{2+}, which has the same size and an extra positive charge to make up for the one lost when Al replaces Si.

coupled substitution

3 Elements do not have to be chemically related to substitute for one another. For example, Fe^{2+} and Mg^{2+} are freely interchangeable in many minerals and substitution of Al^{3+} for Si^{4+} is very common, as we have already noted.

4 With ionic substitution possible on this scale, it is not surprising that the silicates could not be classified by composition alone (Section 2.1). It needed X-ray diffraction to show, for example, that aluminium can be both cation and aluminate anion in the same structure.

5 As a general rule, one ion may replace another if the size difference between substituting and substituted ions does not exceed 15 per cent of the smaller ion.

6 Isomorphism is *not* the same as solid solution.

SAQ 4 Refer to Table 1. Can you explain why fluoride in water or toothpaste can be incorporated into teeth and bone, which are mainly calcium phosphate structures containing $(OH)^{-}$ ions? Could chloride from chlorinated tap water be incorporated as easily?

SAQ 5 The salts KCl and KBr form a complete range of mixed crystals. The salts NaCl and KCl, however, have only limited solid solubility. Explain this, using data from Table 1.

SAQ 6 From your answer to *SAQ 5*, would you expect the two alkali feldspar minerals $KAlSi_3O_8$ and $NaAlSi_3O_8$ to have complete or partial solid solubility – should they form, in other words, a complete or partial solid solution series?

SAQ 7 The olivines Mg_2SiO_4 (forsterite, abbreviated to Fo), and Fe_2SiO_4 (fayalite, abbreviated to Fa) are isomorphous (isostructural). So are the carbonates $CaCO_3$, $MgCO_3$, and $FeCO_3$. The carbonates are also isomorphous with nitre, $NaNO_3$, because both the CO_3^{2-} anion and the NO_3^{-} anion are triangular in shape.

Using available data from Table 1, can you predict whether:
(a) The olivines will form a complete solid solution?
(b) Solid solution will be complete between all three of the carbonates given above?
(c) There can be solid solution between calcite and nitre?

SAQ 8 Is it possible that the two micas muscovite, $KAl_2AlSi_3O_{10}(OH)_2$, and phlogopite, $KMg_3AlSi_3O_{10}(OH)_2$, could form a solid solution series? Give reasons for your answer.

2.5 Polymorphism (*Objectives 1, 4*)

polymorphism

An element or compound with more than one crystal structure is called polymorphic. The phenomenon of polymorphism is well summarized in a section under the same heading, pp. 90–4 in *Principles of Geochemistry*. You should read this and appreciate the following points:

1 Polymorphs have the same composition but different structure.

2 There are two kinds of polymorphic transition, one reversible (enantiotropic), the other non-reversible (monotropic).

3 The rate of polymorphic transition may be slow or rapid, and the stability of a polymorphic form can be influenced by pressure, temperature, and the presence of impurities. Full understanding of the thermodynamic treatment on p. 93 is not essential, but you should understand the implications.

4 There are three types of polymorphic transitions: those involving small displacements only, those in which bonds are broken and remade, and order–disorder transitions.

When you have read the section answer SAQs 9, 10 and 11.

SAQ 9 Distinguish between isomorphism and polymorphism.

SAQ 10 Diamond is the hardest mineral known. It is thermodynamically unstable in the setting of an engagement ring, i.e. at normal atmospheric pressure and temperature. So why can advertisers say 'a diamond is forever'?

Aragonite and calcite are both polymorphs of calcium carbonate, $CaCO_3$. Aragonite is the polymorph precipitated by many marine and non-marine creatures which form protective shells, especially the molluscs, oysters, clams, and so on. But aragonite at the Earth's surface is unstable and inverts to calcite, which is why the calcium carbonate polymorph in limestones is mainly calcite. On the other hand, in some rocks known to have formed at rather deep levels in the crust, under several thousand atmospheres pressure and a few hundred degrees centigrade, aragonite is more commonly found than calcite.

SAQ 11

(a) Using the information on p. 91 in *Principles of Geochemistry*, and the analogy provided by the diamond-graphite transition, draw a semi-quantitative stability graph for the calcite-aragonite transition.

(b) Use this to predict which of the two polymorphs is likely to be the denser one.

(c) If aragonite is unstable at the Earth's surface, how can we possibly *know* that it occurs in rocks formed at substantial depths in the crust?

2.6 Silicate Polytetrahedra (*Objectives 1, 5*)

silicate polytetrahedra

When the ionic and covalent character of the Si–O bond are taken together, we find that it is the strongest bond in the silicates. Thus the silicate oxyanions or tetrahedra $(SiO_4)^{4-}$ form the skeleton of the silicate structures.

oxygen sharing

In all polytetrahedral silicate structures, the tetrahedra are joined by vertices, that is, by sharing oxygen atoms. Any one tetrahedron can share vertices with up to four others, as happens in some structures (quartz and feldspars, for example).

We can draw the silicate anions in a convenient shorthand which can be compared with ball-and-stick models (Fig. 4).

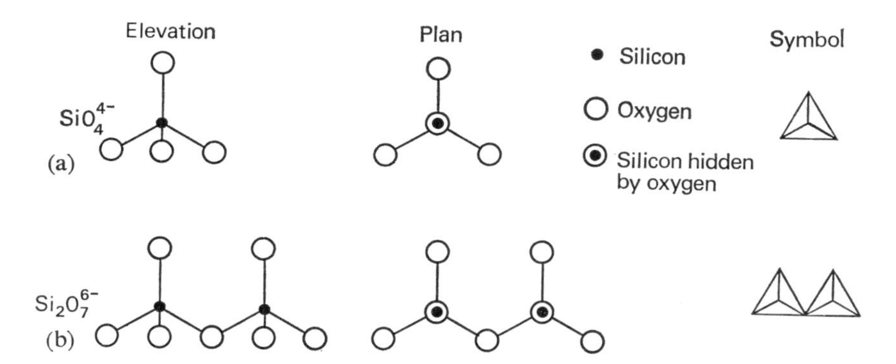

Figure 4 Ball and stick models of silicate structures, and 'shorthand' symbols.

Model Exercise B

You have already constructed a model of the $(SiO_4)^{4-}$ tetrahedron in Model Exercise A. For this you used a plasticine ball with a radius of 4.5 mm (or as near this as you could get!). In relation to the plastic balls, this was approximately the scale size of the silicon ion. For the remainder of the exercises you will use the red and black balls and spokes. So remember that these give no idea of the relative size of the ions we are considering. However, these models do provide an accurate geometric representation of the structures of the important silicate minerals which are considered in this Unit, and we are using them for this reason. The black balls represent silicon ions, the red balls oxygen ions (you will commonly need to use a black ball to represent oxygen, however, and we will note this where necessary.)

Begin by building an $(SiO_4)^{4-}$ tetrahedron and then build an $(Si_2O_7)^{6-}$ group as shown in Figures 4a and b. For the latter you will have to use a black ball to represent the oxygen atom which is shared between the two tetrahedra.
COMPARE MODELS WITH THOSE SHOWN IN FIGURE 33A AND B.

> Given that a single tetrahedron has a net charge of 4–, why does the Si_2O_7 configuration, which is two tetrahedra together, have a net charge of only 6–?

The simplest way to explain this is to examine Figure 5a which shows how the charges become distributed on the tetrahedron.

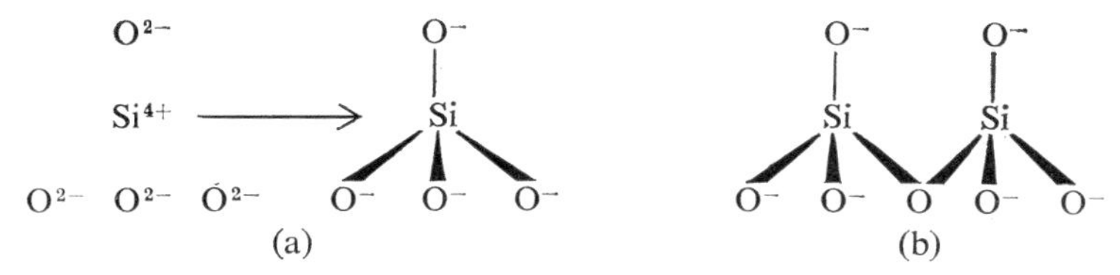

Figure 5 (a) Four oxygens combining with a silicon (Si^{4+}) give the tetrahedral structure with four surplus negative charges, one on each oxygen. (b) Linking tetrahedra by oxygen-sharing results in the shared oxygen becoming neutral.

When two tetrahedra combine together by sharing an oxygen at one or other tetrahedral vertex, *the shared oxygen is neutral, because it is linked to two silicons* (Fig. 5b). This is an invariable rule in silicate polytetrahedra and is a very important principle to grasp. To show that you have grasped it, do Model Exercise C and answer *SAQ 12*.

Model Exercise C

Build the polytetrahedral arrangement symbolically shown in Figure 6. *Note*: you will need to use black balls for the shared oxygens as well as for the silicons.
COMPARE YOUR MODEL WITH THAT SHOWN IN FIGURE 33C.

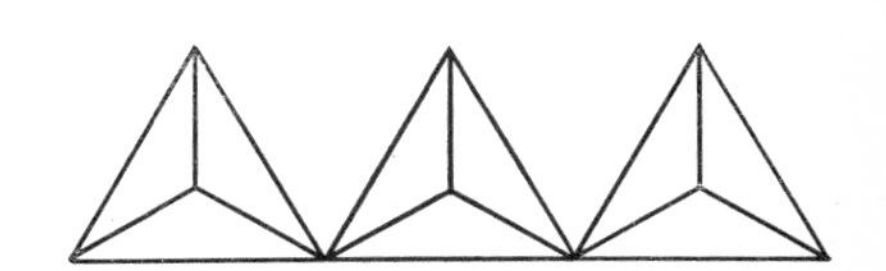

Figure 6 Silicate structure formed by the linking of three SiO_4 tetrahedra.

SAQ 12 What would be (a) the formula of, (b) the net charge on, the polytetrahedral arrangement you have made?

Remember that the negatively charged tetrahedra and polytetrahedra which form the framework of silicate mineral structures are held together and neutralized by positively charged metal cations, to make very stable configurations.

Pauling's rules. Linus Pauling* has compiled five rules for stable ionic structures, such as those of the silicates. They are given in Appendix 4 (Black), which you could read at once or after the description of silicate structures, which follows now.

2.7 Structural Classification of the Silicates *(Objectives 1, 4, 5, 6, 7)*

A useful summary of this Section is on pp. 81–3 in *Principles of Geochemistry*, with a thermodynamic supplement (black page) on pp. 83–6. Read the summary before proceeding with this Section and again after you have finished it.

2.7.1 Separate tetrahedra – the orthosilicates SiO_4^{4-} groups

You constructed a single SiO_4^{4-} anionic group in Model Exercise B. Minerals which contain such separate silicate anions are the *olivines* (Mg, Fe)$_2$ SiO_4, common in basic igneous rocks and believed to be the main constituent of the upper mantle (Plate A,1). The name comes from their colour which is olive green due to ferrous iron. The series is a good example of solid solutions formed by substitution (Section 2.4) of Mg^{2+} and Fe^{2+}. There is no substitution of silicate by aluminate in the olivines. Figure 7 is a projection of the idealized structure of forsterite, Mg_2SiO_4, as determined by X-ray diffraction. This is a very compact structure indeed. The oxygens are very nearly hexagonally close-packed, with the silicons in tetrahedral holes (obviously!), and Mg^{2+} and Fe^{2+} in octahedral holes (Section 2.3.1). The red circle in Figure 7 shows an octahedron of oxygens around Mg^{2+}.

orthosilicate

olivine structure

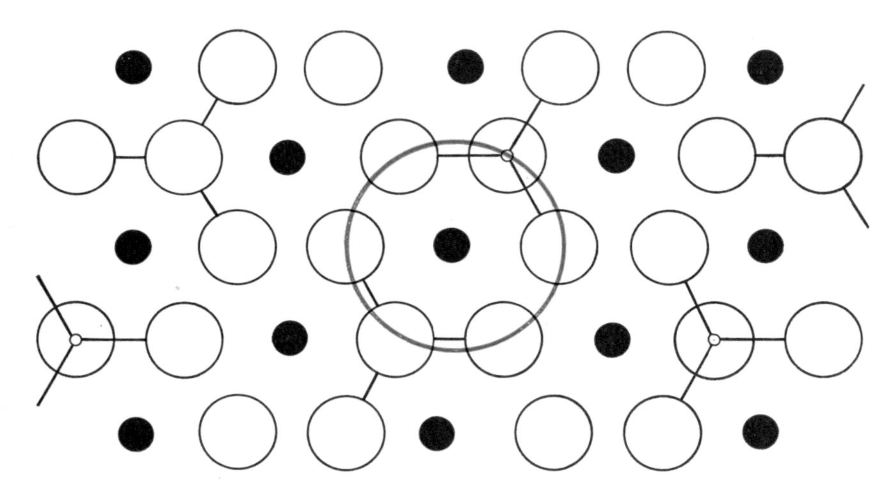

Figure 7 The atomic structure of the magnesium olivine, forsterite. (Large circles – oxygen; small circles – silicons; black – magnesiums.) Two rows of SiO_4 tetrahedra are shown with their oxygens in planes parallel to the paper. Each tetrahedron is associated with three metallic cations (Mg^{2+}): these three are alternately behind tetrahedra pointing up and in front of tetrahedra pointing down. The red circle is drawn through an octahedron of oxygens around Mg^{2+}.

The *garnets* (Plate A,2) form another group of important minerals with general formula $M_3^{2+}M_2^{3+}(SiO_4)_3$. They also contain separate tetrahedra, but have a more complex structure and varied composition than the olivines. The compact

* *Nobel Prize for Chemistry, 1954; Nobel Prize for Peace, 1964.*

structure of the olivines and other silicates with separate tetrahedra results in a high density. This means that they are relatively stable at high pressures, under conditions characterizing the upper mantle (Section 2.14) and the lower crust (Section 2.13).

2.7.2 Double tetrahedra $Si_2O_7^{6-}$ groups

This is a small class of rare minerals. It includes a zinc mineral, hemimorphite, with a bent Si–O–Si bond (Fig. 8a) and a scandium mineral, thortveitite, with a linear Si–O–Si bond (Fig. 8b).

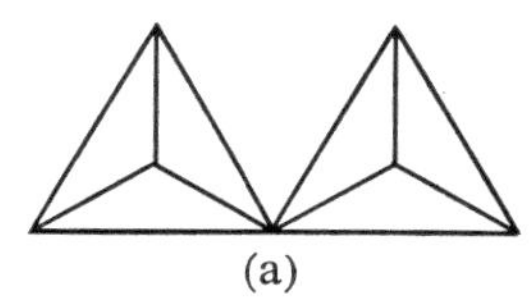

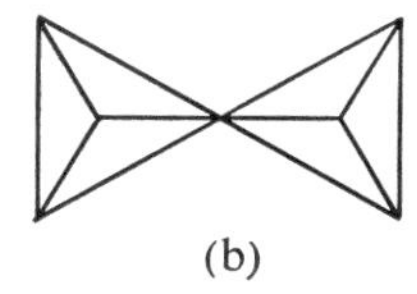

(a) (b)

Figure 8 (a) Arrangement of SiO_4 tetrahedra in hemimorphite. (b) Arrangement of SiO_4 tetrahedra in thortveitite.

2.7.3 Ring structures – metasilicates or cyclosilicates $(SiO_3)_n^{2n-}$ groups

These minerals are also rare and may contain 3, 4, or (most commonly) 6 tetrahedra (Fig. 9). An example of the latter is beryl, a beryllium ore (containing also Al^{3+}), some varieties of which have long been used as semi-precious stones (Plate A,3). The colours are due to impurities, the deep green of emerald being due to trace amounts of Cr. The structure of beryl is shown in Figure 10.

ring silicate

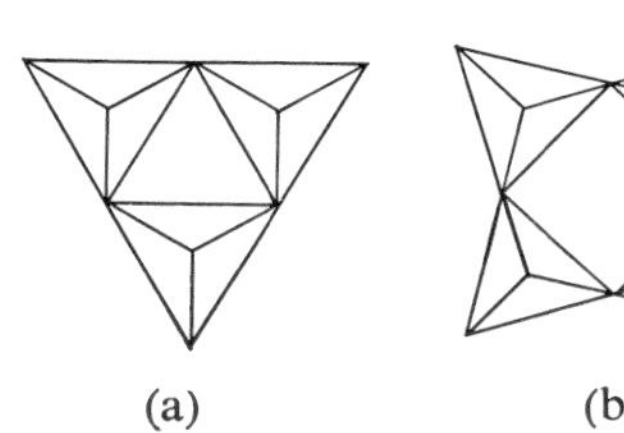

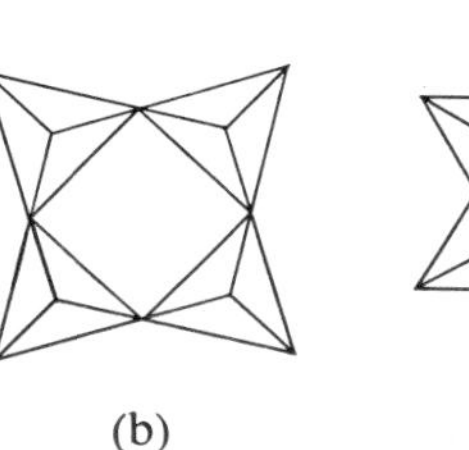

(a) (b) (c)

Figure 9 Arrangement of SiO_4 tetrahedra in ring silicates: (a) with 3 tetrahedra, e.g. benitoite ($BaTiSi_3O_9$); (b) with 4 tetrahedra, e.g. axinite; (c) with 6 tetrahedra, e.g. tourmaline, beryl.

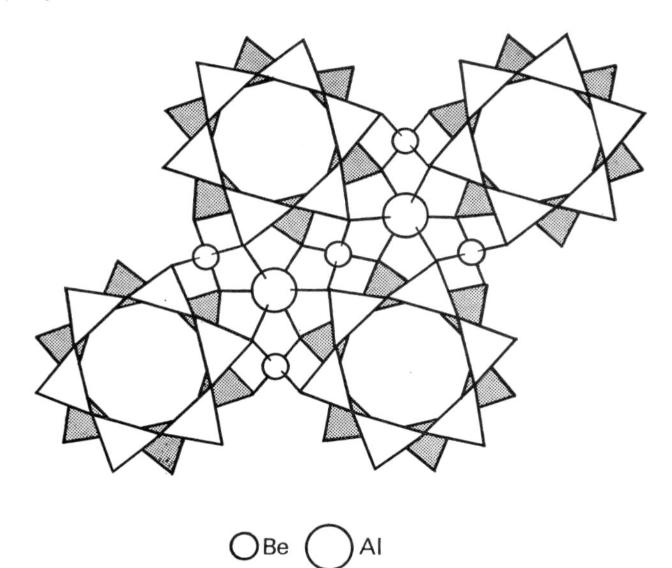

Be Al

Figure 10 The structure of beryl, a ring silicate which has rings containing six SiO_4 tetrahedra. The $(SiO_3)_6^{12-}$ rings are at two different heights, and are represented in this projection as rings of shaded or unshaded tetrahedra. The rings at different heights are staggered, and Be^{2+} and Al^{3+} lie in a plane midway between them. The oxygens inside the rings are all shared and therefore neutral, so there are no cations inside the rings.

Model Exercise D

Make models of a 4-membered ring (Fig. 9b) and a 6-membered ring (Fig. 9c). You can use red balls for all oxygens. What are the formulae of these configurations?

COMPARE YOUR MODELS WITH THOSE SHOWN IN FIGURE 33D AND E.

Answers $=(Si_4O_{12})^{8-}$

$(Si_6O_{18})^{12-}$

The rather more common black mineral, tourmaline, found especially among the mining districts of south-west England, is another ring silicate with a fairly complex chemical composition.

The silicate structures that follow all contain giant anions (S100[8]) extending indefinitely in one, two or three dimensions.

2.7.4 Single chains – the pyroxenes $(SiO_3)_n^{2n-}$

chain silicate

In pyroxenes, two vertices of each tetrahedron are shared. The repeating unit (indicated by the double-headed arrow in Figure 11) is two tetrahedra, $(Si_2O_6)^{4-}$. Cations bind the chains together, as in *diopside* $CaMg(SiO_3)_2$. The commonest pyroxene in rocks is *augite*, which resembles diopside, but contains also Fe and Al (Plate A, 4).

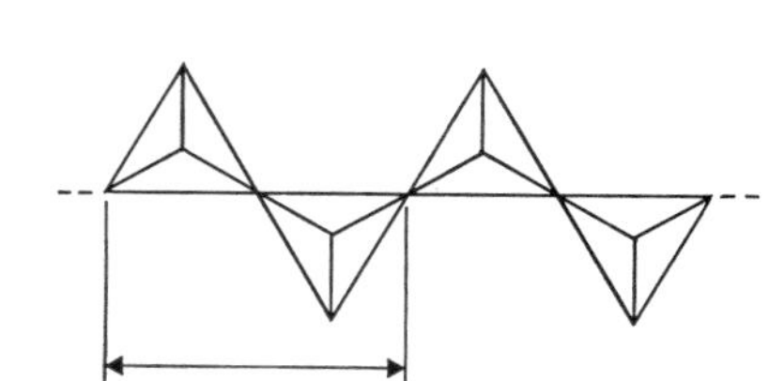

Figure 11 Arrangement of SiO_4 tetrahedra in pyroxenes.

Model Exercise E

Make a model to represent a portion of the silicate polyanion group structure in a pyroxene chain as illustrated in Figure 11. You can extend it in one dimension as much as you like – the only limitation is the size of your kit – because this is the first anion of indefinite dimensions that we have considered. All the structures which you will build up from now on will be of the same type.

COMPARE YOUR MODEL WITH THAT SHOWN IN FIGURE 33F.

Pyroxenes are common in basic igneous and metamorphic rocks. They are fairly dense and hard, but less so than olivines, and are therefore less important in the upper mantle. Unlike olivines, they possess a characteristic plane of weakness or *cleavage*, parallel to the chains (in planes at 93° and 87° to each other), along which crystals preferentially break or split (Fig. 12). This is because the ionic bonds holding different chains together are not as strong as the Si–O bonds within the chains.

cleavage

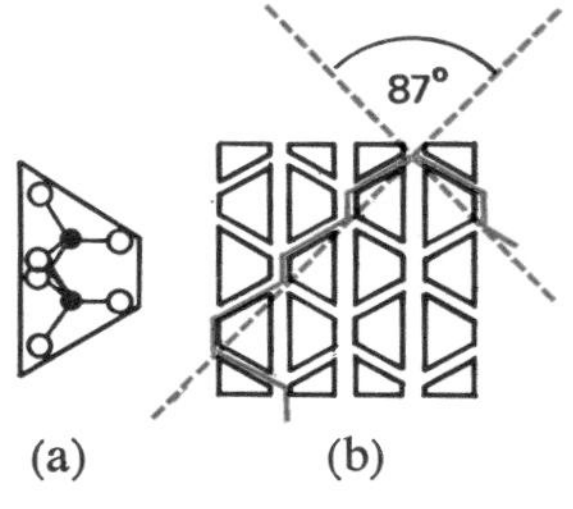

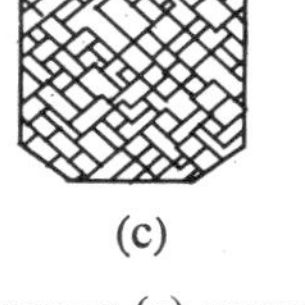

Figure 12 Relationship between cleavage and structure in pyroxenes: (a) arrangement of tetrahedra as seen looking lengthways down a pyroxene chain; (b) relations of cleavage (broken lines) to chains, looking lengthways along part of a pyroxene crystal made of many parallel chains; (c) appearance of cleavages in a crystal cut normal to the chains.

A rare pyroxene, jadeite, $NaAl(Si_2O_6)$, may be familiar as a constituent of the ornamental stone, jade. It is characteristic of rocks found at high pressures and this is reflected in the octahedral coordination of its Al (noted in Section 2.3). Jadeite is the characteristic pyroxene of eclogite, a high pressure equivalent of basalt, which you will meet later in the Unit.

2.7.5 Double chains or bands – the amphiboles $(Si_4O_{11})_n^{6n-}$

Each double chain or band consists of a *pair* of linked pyroxene-type chains. Two and three vertices alternatively are shared, and the tetrahedra all point out of the plane of the paper in Figure 13.

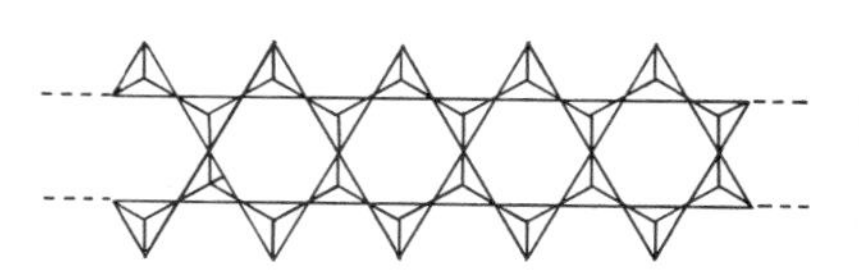

Figure 13 Arrangement of SiO_4 tetrahedra in the amphibole structure.

Model Exercise F

Make a model to represent a portion of the silicate polyanion structure in an amphibole, as illustrated in Figure 13.

band silicate

COMPARE YOUR MODEL WITH THAT SHOWN IN FIGURE 33G.

There is a series of hexagonal holes in each band, each hole containing a hydroxide ion, $(OH)^-$, which can be replaced by F^-. Some silicate tetrahedra can be replaced by aluminate. The cations (Ca, Mg, etc.) neutralize not only the unshared oxygens on the tetrahedra, but also the hydroxide ions and the excess charges due to Al^{3+} substituting for Si^{4+} in the tetrahedra. The cations hold the bands together rather less strongly than in pyroxene, for amphiboles show a more pronounced cleavage. This has a distinctive angle of 56° and 124°, as opposed to 87° and 93° for pyroxenes (Fig. 14).

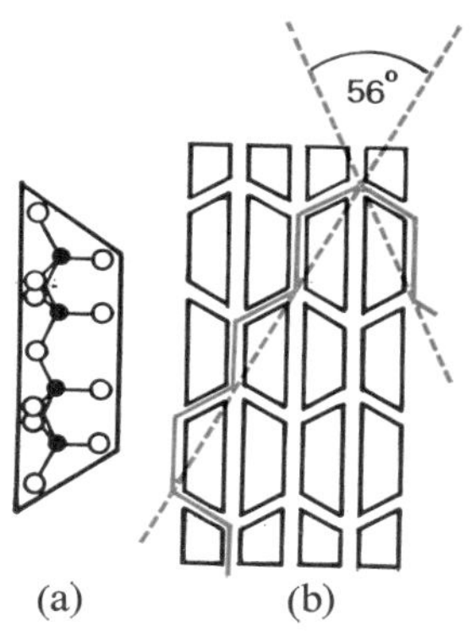

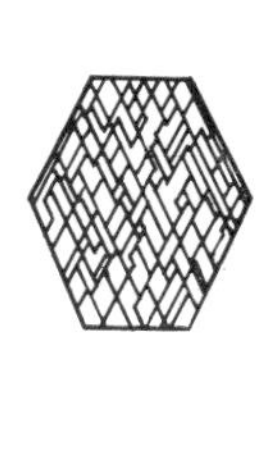

Figure 14 Relationship between cleavage and structure in amphiboles: (a) arrangement of tetrahedra as seen looking lengthways down an amphibole band; (b) relation of cleavage (broken lines) to chains, looking lengthways along part of an amphibole crystal made of many parallel bands; (c) appearance of cleavages in a crystal cut normal to the chains.

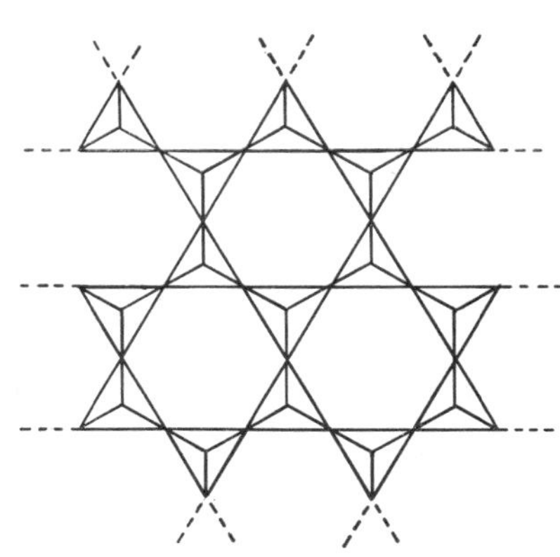

Figure 15 Arrangement of tetrahedra in a sheet silicate structure.

An example of a simple amphibole is *tremolite*, $Ca_2Mg_5(Si_8O_{22})(OH)_2$. Some tremolite occurs in a very fibrous form, the fibres reflecting the band structure. This type is a variety of asbestos. The fibres can be twisted to form string, or woven into cloth and, since it does not melt readily, it can be used for safety curtains and protective clothing. It is a thermal insulator because of the air pockets between the fibres.

The most important amphibole in rocks, however, is dark green hornblende which forms long crystals, but not fibres. This contains some Fe^{2+} replacing Mg^{2+}, and some Al^{3+} replacing Si^{4+}.

2.7.6 Layer or sheet silicates $(Si_4O_{10})_n^{4n-}$

Three vertices of each tetrahedron are shared, to form two-dimensional giant anions which are potentially infinite sheets of honeycomb pattern (Fig. 15), the crystal form often reflecting the hexagonal symmetry within the sheets. The layer structures are many and varied. We will describe first of all *talc* and *mica*, to show how the physical properties depend on the organization of the layers, and later on (in Section 2.9) the *clays*, because of their general importance (Unit 1, Section 1.5).

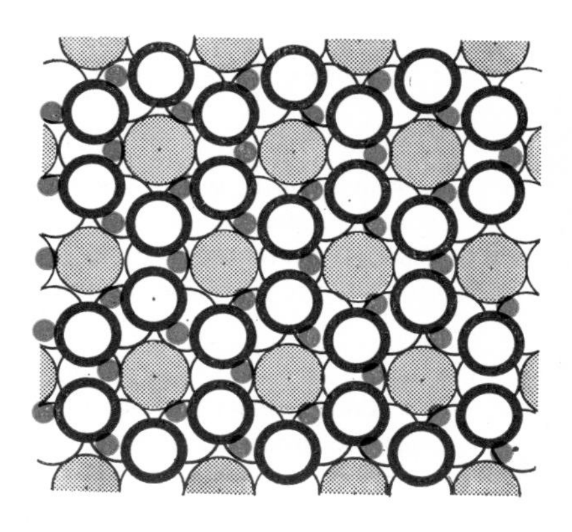

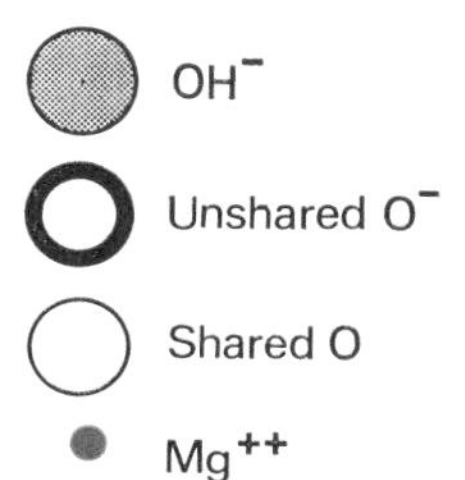

Figure 16 The bottom layer and inside of the talc sandwich (plan view).

Model Exercise G

Take your last model and (partially dismantling it, if necessary) add to it so as to make it into a sheet such as that illustrated in Figure 15. This represents the silicate polyanion structure in a mica sheet. Note that you could indefinitely extend this structure in two dimensions and not in just one as with the chain and band silicate models you made in Model Exercises E and F. But if you make a single ring of hexagonal shape, you have part of the beryl structure (Fig. 9c) as in Model Exercise D.

COMPARE YOUR MODEL WITH THAT SHOWN IN FIGURE 33H.

Talc

Talc, or soapstone, a magnesium silicate, $Mg_3Si_4O_{10}(OH)_5$, is one of the softest minerals known. The structure consists of double layers like sandwiches. The bottom 'slice of bread' is a silicate sheet, as in Figure 15, with a hydroxide ion in each of the hexagonal holes formed by the unshared oxygens at the apex of each tetrahedron, as in Figure 16 (and as in the amphiboles). The result is effectively a close-packed layer of singly charged oxygens, because the $(OH)^-$ ions resemble unshared O^- (p. 76, *Principles of Geochemistry*, and Table 1 of this Unit). Sitting on this layer, like the filling in a sandwich, are the magnesium cations that hold the bottom and top layers of the sandwich very firmly together. The top layer is a mirror image of the bottom one, but slightly displaced (Fig. 17) so that the two inner layers of oxygen form a close-packed array (compare Fig. 3c, d), with magnesium in octahedral holes. Figure 16 shows all the magnesium in the sandwich, but only the bottom sheet of oxygen and OH^-, as if we had taken the top slice off.

talc structure

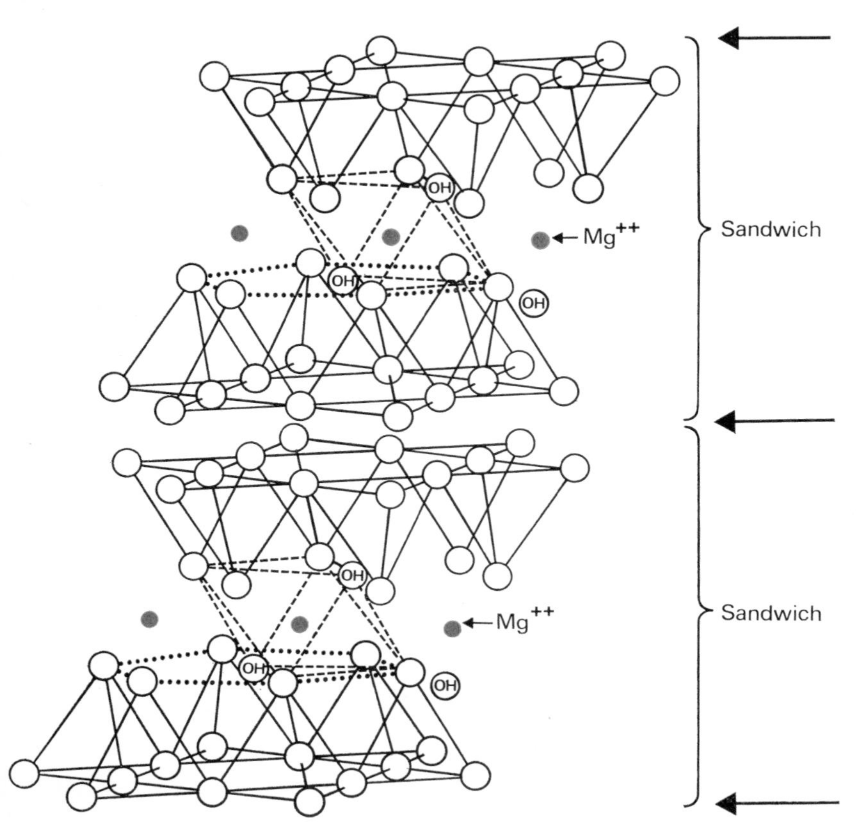

Figure 17 The talc sandwich viewed sideways on. (Compare with Fig. 16 for plan view.) The silicons inside the SiO_4 tetrahedra are not shown. The dotted hexagons pick out the six oxygens (forming the 'hexagonal hole') around OH^-. The dashed octahedra show the 6-fold coordination of Mg^{2+} by four oxygen atoms (two from each sheet) and two (OH^-) ions (one from each sheet). Large arrows show planes of sliding or cleavage.

The remarkable property of this structure is that all the anionic charges are *inside* the sandwich, or double layer. The tetrahedra all point into the sandwich, and on these only the unshared oxygens are charged. These charges, and those on the OH^- ions, are neutralized by the Mg^{2+} in the sandwich.

What holds the sandwiches together, then, to make crystals of talc?

There are no ionic or covalent forces to bind a stack of such layers together. All that remain are residual or van der Waals forces†, which are very weak, so the layers can readily slide across each other. The large arrows in Figure 17 show the plane along which the non-bonded surfaces of each layer meet and along which they can slide freely.

The structure elegantly explains the unusual qualities of talc. It explains the well-developed cleavage and free sliding which occurs between the weakly bound layers, which, in turn, explains why talc has a greasy feel and is a good lubricant (as in talcum powder), and why one can write with it, as with graphite, which is also soft and has a layer structure (S100[9]). The mineral *serpentine* is related to talc, but we shall return to this more appropriately in Section 2.9.

The micas

mica structure

The micas, such as phlogopite, $KMg_3(AlSi_3O_{10})(OH)_2$, are closely related to talc. Mica has a pronounced layer cleavage, as you remember from S100[10], and the layers have the same hexagonal structure as do those of talc. But whereas a thin sheet of talc can be readily folded and unfolded, thin mica sheets are elastic – they do not stay folded, but spring back unless broken. In fact, mica can be cleaved to give strong transparent sheets of less than 1/100 mm thickness. Again, the crystal chemistry elegantly explains these resemblances and differences. Mica and talc have the same double-layer structure, with the unshared oxygens of the tetrahedra pointing inwards into the sandwich and interspersed with OH^- ions in the hexagonal holes of the silicate layer. Phlogopite mica even has magnesium ions inside the sandwich holding it together, as does talc, and Figure 16 can also serve as a picture of the inside of a phlogopite mica sandwich. But the difference between mica and talc arises from the presence of *aluminate tetrahedra among the silicates in mica.* One in four of the mica tetrahedra is aluminate $(AlO_4)^{5-}$, as opposed to silicate $(SiO_4)^{4-}$.

What does that mean in terms of net charges on the outer layers of each sandwich?

In talc, unshared oxygens on the $(SiO_4)^{4-}$ tetrahedra and $(OH)^-$ ions were all balanced by magnesium in the sandwich filling (Fig. 17). But one in four tetrahedra now has an extra single negative charge, and the number of magnesium ions remains the same.

What must happen?

The outer surfaces of the sandwiches can no longer be neutral, but must be negatively charged. The extra *positive* charge required to neutralize this is in the form of potassium ions in the planes between each sandwich and the next. These are shown in Figure 18, which is much the same as Figure 17, except for the K^+ ions between the sandwiches in mica and the presence of aluminate among the silicate. Because the net negative charge is rather low (contributed from only one in four tetrahedra remember), a large positive ion is needed, which is why potassium is found here (Pauling's Rule 2, Appendix 4 (Black)).

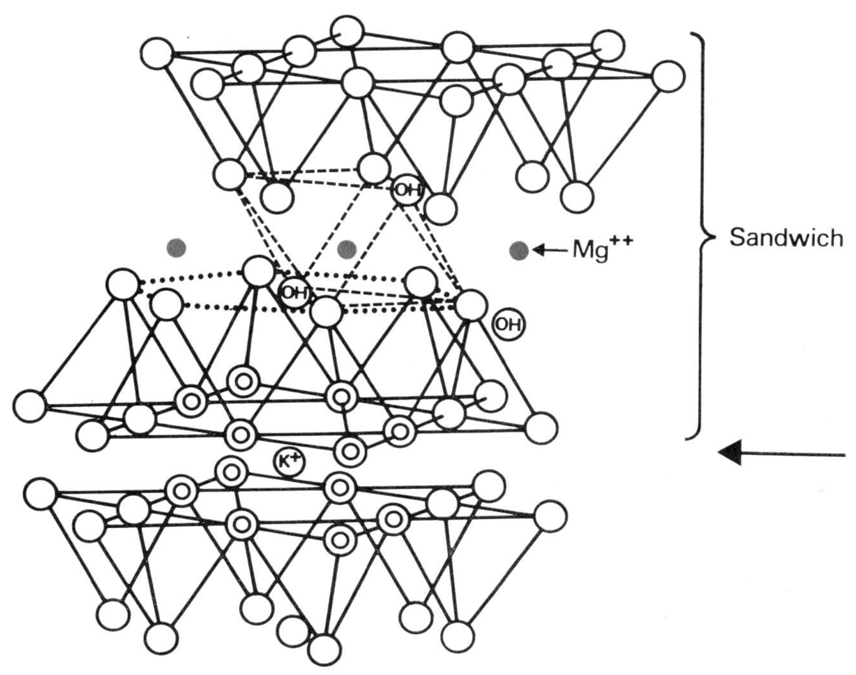

Figure 18 The mica sandwich viewed sideways on (cf. the structure of talc, Fig. 17). Note that each K^+ lies at the mid-point of a line joining the centres of two hexagons, each of which is formed by the basal oxygens of the tetrahedra forming the layers. Large arrow shows plane of cleavage.

Figure 18 shows that the K^+ ions are 12-coordinate, which is just possible with ions of this size (Table 4.2 in *Principles of Geochemistry*).

The effect of aluminate substitution upon the physical properties of mica as compared with talc are now plain. In mica, the layers or sheets will not slide against each other, because the potassium ions are holding them together. Rather than stay folded, mica layers will spring back or, if bent far enough, will snap. All the same, the ionic bonding between one layer and the next is very weak, since the potassium is only singly charged and 12-coordinate. So there is still a very well developed layer cleavage (along the large arrow in Fig. 18).

There is quite a range of micas with different cations between the layers. The commonest mica in granitic rocks is black *biotite* which has Fe^{2+} substituted for the Mg^{2+} of phlogopite. Colourless *muscovite* mica (Plate A, 5) has octahedral Al^{3+} replacing the Mg^{2+} within the sandwiches: $2Al^{3+}$ replaces $3Mg^{2+}$ so that the net charge in the alumino-silicate is unchanged. Muscovite is thus an example of a structure in which aluminium plays two different roles, as cation (6-coordinate by oxygen) and as aluminate (4-coordinate by oxygen). In rare cases Cr^{3+} may replace some Al^{3+} in the 6-coordination positions, yielding the bright green chromium muscovite, *fuchsite*. Because of its high reflectivity, ground muscovite is used as a filler for paper and paints. The heat resistance and low thermal conductivity of muscovite ensure the continued use of large crystals as windows for furnaces and for support of the heating elements in electric irons and toasters.

The clay minerals should properly be treated here, but we shall defer consideration of them to Section 2.9, for reasons which will become apparent then.

2.7.7 Framework silicates

In *framework silicates*, all four of the vertices on each tetrahedron are shared, to form three dimensional framework structures.

framework silicate

SAQ 13 Explain how sharing of oxygen between SiO_4 tetrahedra can make a neutral structure with the formula $(SiO_2)_n$.

The silica minerals

The framework structure made entirely by sharing of SiO_4 tetrahedra is the covalent giant molecule $(SiO_2)_n$, silica, the commonest form of which is *quartz* (Plate A, 6) (S100[8]).

silica structure

At the beginning of this Unit (Section 2.1) we spoke of fruitless early attempts to characterize silicates by reference to various kinds of silicic acids. In fact, some silica minerals can be formed from orthosilicic acid, H_4SiO_4. An aqueous solution

PLATE A

Some silicate mineral crystals.

1 *Olivine, $(Mg, Fe)_2SiO_4$.*

2 *Garnet (almandine, $Fe_3Al_2(SiO_4)_3$), reddish crystals in rock matrix.*

3 *Beryl ($Be_3Al_2Si_6O_{18}$), variety aquamarine, six-sided crystals.*

4 *Pyroxene (augite, $Ca(Fe,Mg)(SiO_3)_2$), short stumpy crystal in matrix.*

5 *Mica (muscovite, $KAl_2AlSi_3O_{10}(OH)_2$), tabular crystals in matrix.*

6 *Quartz, SiO_2.*

(Photographs 3 and 5 are by courtesy of the Photographic Department, Institute of Geological Sciences. The rest were taken by the Open University Photographic Section, from samples supplied by the Mineralogical Section, Institute of Geological Sciences.)

1

2

3

4

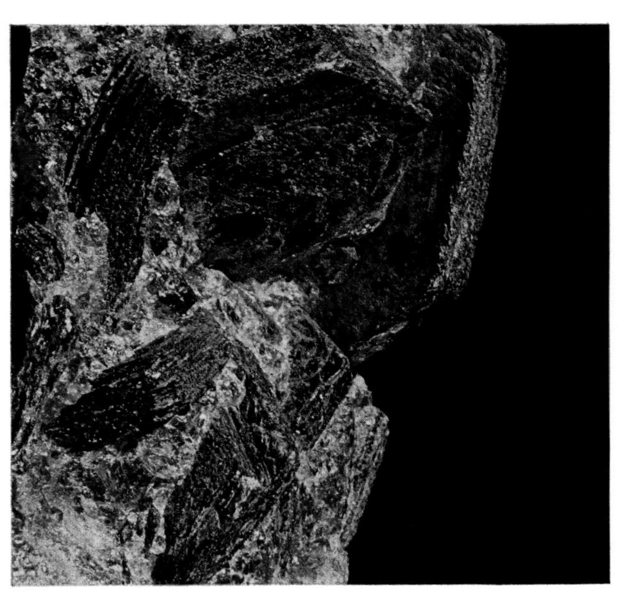
5

6

1

2

3

4

5

6

In S100[17] we did not have the space to elaborate on our reasons for supposing that the upper mantle is composed of peridotite. We shall do this now as an example of just how much geochemical information it is possible to obtain about a physically inaccessible region like the upper mantle. Our argument runs as follows, although of course it was not originally put together in this neat form:

1 P wave velocities for the upper mantle are in the order of 8.1–8.3 km s^{-1}. Table 8 is a list of different kinds of rocks with their seismic velocities, as measured in the laboratory.

Table 8

Rock type	*P wave velocity (km s^{-1}) at pressures corresponding to upper mantle*
greywacke	6.20
slate	6.22
granite	6.45
granodiorite	6.56
gabbro	7.24
peridotite	8.15

Clearly, of these alternatives, peridotite is the best candidate.

2 Because it is made of a mixture of minerals, peridotite does not have a single sharp melting point but instead melts over a *range* of temperatures. Laboratory experiments with natural and artificial peridotite samples have shown that when these begin to melt, the first 10–20 per cent of liquid produced has the chemical composition of basalt. (The process of partial melting is discussed in greater detail in Unit 4.)

3 From seismic measurements we know that the earthquakes which precede basaltic lava eruptions commonly have foci (S100[18]) at depths of 60 km or more, well within the upper mantle. Since the earthquakes are likely to be occurring at the site of basalt magma generation within the mantle, we infer that basaltic lavas originate in the upper mantle.

Put together, these three pieces of evidence suggest that the upper mantle consists of peridotite which locally undergoes partial melting to yield basaltic liquids. These rise to the surface and are erupted as basalt lava flows.

Until recently, however, there was another theory for the composition of the mantle, namely that the mantle was composed of eclogite, the dense rock composed of magnesium-rich garnet and jade-like pyroxene and having the chemical composition of basalt, which we have already spoken of in Section 2.13, and which also has a P wave velocity of around 8 km sec^{-1}. The eclogite mantle model has now been largely discarded, for several reasons.

Four of these reasons were outlined to you in S100[17]. The most important objection described there was that pressures equivalent to a depth of 60 km beneath normal continental crust are needed to convert basalt to eclogite – clearly, this is in direct conflict with the results of experiments by Ringwood and Green (Section 2.13) although, as we have said, these may themselves not be entirely valid (Section 2.13.1).

Whatever conclusions one may draw from the discussions in Section 2.13 and 2.13.1, it is clear that basalt will convert to eclogite at depths of more than 30 km and that the oceanic crust, which covers 70 per cent of the Earth, is nowhere more than 10 km thick and commonly a good deal less. Such thicknesses are clearly quite insufficient to maintain an eclogitic upper mantle beneath the ocean floor. Moreover, the Moho is quite as distinct beneath the oceans as under the continents, and this is more consistent with a fundamental change in composition than with a phase change (Section 2.5), which would probably take place over a somewhat diffuse zone rather than across a distinct boundary.

Since we have no direct access to the upper mantle, it may seem rather paradoxical that we have pieces of peridotite which we believe to have come from there. How has this happened?

2.14.1 Ultramafic† nodules

One of the most exciting and interesting features of many volcanoes is the rich variety of inclusions which their lavas contain—inclusions of rocks torn off deeper Earth layers by the lava as it passes through them. Volcanoes are thus natural 'drill-holes' which provide us with samples of the Earth from various depths. Many of these inclusions are of common crustal rocks such as granite and granodiorite – and intermediate granulites, incidentally, which is at least *consistent* with the granulite model for the lower crust (Section 2.13), even if it does nothing to *prove* it! Others, rather more rare but found all over the world (both on continents and in the ocean), are of peridotite and occur as rounded blocks (nodules) ranging from below 1 to over 20 cm across (we have some fine examples in Derbyshire and on the Fifeshire coast). Broadly speaking there are two main types:

ultramafic nodules

Olivine nodules (Plate B, 1–4) are so called because they are very rich (over 80 per cent) in olivine. Their mineral composition is surprisingly consistent, with only small variation in the proportion of minerals present. Four minerals are always found. These are listed in Table 9, in order of relative abundance.

olivine nodules

Table 9 The major minerals of olivine nodules

Mineral	*Formula*	*Structure*
olivine	$(Mg, Fe)_2SiO_4$	olivine
diopside	$CaMgSi_2O_6$	pyroxene
enstatite	$MgSiO_3$	pyroxene
chrome spinel	$(Fe,Mg)O(Al,Cr)_2O_3$	oxide (non-silicate)

The mineral formulae are somewhat simplified. For example, the pyroxenes contain significant amounts of Fe^{2+}, Al^{3+}, Cr^{3+} and Na^{+}, and the olivine contains significant amounts of Ni^{2+}.

The only evidence we have that these nodules come direct from the mantle is circumstantial – we believe that the mantle is made of peridotite, and here are pieces of peridotite in basaltic lavas which we know to have originated in the mantle. Such arguments are clearly in danger of becoming circular! Indeed there is a robust school of thought which holds that olivine nodules are not mantle materials at all, but have, in fact, formed by accumulation of minerals crystallized from the basalts, a process called crystal fractionation, which we examine in Unit 4. Their case cannot be completely disproved, although they are in a minority at present.

Garnet peridotite inclusions, however, are much more widely accepted as being of upper mantle origin. They are more variable in composition and carry, in addition to the minerals listed above, others such as *pyrope*, the magnesium garnet ($Mg_3Al_2Si_3O_{12}$) and *phlogopite*, the magnesium mica ($KMg_3AlSi_3O_{10}(OH)_2$, Section 2.7.6). These inclusions are chiefly remarkable, however, because they rarely occur in ordinary basalts but are found, along with diamonds, in *kimberlite*. Kimberlites are rare rocks which occur in steep-sided pipes formed by explosive volcanism and called *diatremes*. They are named after famous occurrences at Kimberley in South Africa. Kimberlite is an ultrabasic igneous rock, containing fragments of garnet peridotite and eclogite and of minerals such as garnet and diamond (Plate B, 6), which formed under high pressure. These are set in a fine-grained matrix which contains calcite and micas – these respectively contain CO_2 and H_2O, and are believed to reflect a high gas pressure during the formation of the rock. Because kimberlites contain broken fragments of rock and show evidence of considerable gas pressure, they are thought to have been explosively intruded. The presence of diamond provides a valuable clue to the depth of origin of kimberlites.

garnet peridotite

kimberlite

You have already studied polymorphism, using calcite/aragonite and diamond/

graphite as examples (Section 2.5). We shall now look at an application of the polymorphism of carbon to a geological problem. If you are unsure of the reason why diamond is the high pressure polymorph of carbon, look back to Section 2.5 or S100[19].

You have already seen a graph showing the fields of stability of diamond and graphite (Fig. 4.4 in *Principles of Geochemistry*). By comparing this with the geothermal gradient† it is possible to determine the minimum depth within the Earth at which diamond is stable.

SAQ 22 Look at Figure 4.4 in *Principles of Geochemistry*. Draw a line on this corresponding to a geothermal gradient of 25 °C per kilobar, noting that the temperature scale is labelled in K. Start the geothermal gradient at 0kb (1 atm pressure is only 1/1000 kb) and 300 K (a hot room temperature!). Now, what is the minimum depth at which diamond can be formed within the Earth?

1 0 km

2 About 70 km

3 About 150 km

4 About 220 km

Clearly, the minimum depth is well below the Moho, which means that kimberlites come from deep in the mantle, much deeper than most basalts. That is one reason why garnet peridotite nodules are widely accepted as having originated in the upper mantle. Another reason is that the minerals of garnet peridotite could not easily crystallize from magmas of kimberlite composition.

2.14.2 Alpine-type peridotites

Alpine-type peridotite

In mountain ranges, vast 'pods' (lenticular masses) of peridotite are found emplaced within the tightly folded sedimentary rocks which formerly accumulated beneath the sea. The enormous stresses involved in mountain building are supposed by many to have thrust great pieces of mantle (with dimensions measureable in km) into upper parts of the crust. These rocks are found in many fold mountains, but they are called Alpine-type because they were first recognized in the Alps.

Their composition is mineralogically similar to that of the olivine nodules we have just described. Again, we are in danger of circular argument in saying these are pieces of upper mantle – and there is an alternative school of thought (again a minority) which holds that, like olivine nodules, Alpine-type peridotites are formed well within the crust, by gravity settling of olivine crystals from basaltic magma (see earlier comment in Section 2.14.1).

We should also mention here that great sub-horizontal slabs of ultramafic (peridotitic) rocks are found in some places, perhaps the best studied being those of the Troodos Mountains of Cyprus and the Oman Peninsula. The slabs of peridotite are overlain by basic rocks arranged very much like the cross-section of Figure 21, and this has led to the plausible suggestion that these slabs are pieces of upper mantle and overlying oceanic crust which have been thrust on to the surface by the powerful Earth movements characteristic of destructive plate margins.

2.14.3 Chemical composition of the upper mantle

Even those who are unimpressed by arguments suggesting olivine nodules and Alpine-type peridotites to be upper mantle fragments, however, are in agreement that the upper mantle is of peridotite composition. What we shall do now is use the majority view and help you to make some deductions about upper mantle compositions from the chemical data available. We present some analyses of the three main groups of rocks which we believe to represent upper mantle materials, together with an analysis of basalt (Table 10).

Table 10 Chemical analyses of rock samples believed to have come from the mantle, and an analysis of basalt

	Peridotites			*Basalt*
	1	2	3	4
SiO_2	41.32	43.4	45.7	49.0
TiO_2	<0.1	0.08	0.13	1.0
Al_2O_3	0.54	1.8	2.7	18.2
Fe_2O_3	1.21	2.7	1.6	3.2
FeO	5.91	6.7	5.7	6.0
MnO	0.11	0.16	0.12	0.2
MgO	49.81	42.4	41.5	7.6
CaO	<0.1	1.8	2.0	11.2
Na_2O	0.05	0.21	0.22	2.6
K_2O	0.005	0.01	0.03	0.9

1 Average Alpine-type peridotite.
2 Average olivine nodule (27 analyses).
3 Average garnet peridotite (3 analyses).
4 Average basalt (Unit 1, Table 3).

You can use the data in Table 10 to make some suggestions about the chemical composition of the mantle by answering the following questions.

SAQ 23 Some elements are *very much higher* in abundance in basalt than in any peridotite. These elements are critical in any consideration of how much basalt can be generated by partial melting of peridotite. Tick the appropriate boxes below to indicate which they are.

SiO_2	TiO_2	Al_2O_3	Fe_2O_3+FeO	MnO	CaO	Na_2O	K_2O
☐	☐	☐	☐	☐	☐	☐	☐

This reflects a very important principle—one which can be used to assess the suitability of different peridotite types as source material for basalt. You can see this by answering *SAQ 24*.

SAQ 24

(a) How much of the Alpine-type peridotite (column 1) containing 0.005 per cent K_2O would have to be melted to give a basalt (column 4) with 0.9 per cent K_2O?
Express your answer as a percentage of the original peridotite and assume that all the potassium goes into the basalt.

(b) How much of the garnet peridotite (containing 0.03 per cent K_2O) would have to be melted to get the same basalt?

(c) By how much is your answer to (b) greater or less than your answer to (a)?

(d) Hence, which of the two peridotite types seems more suitable as a source of basalt?

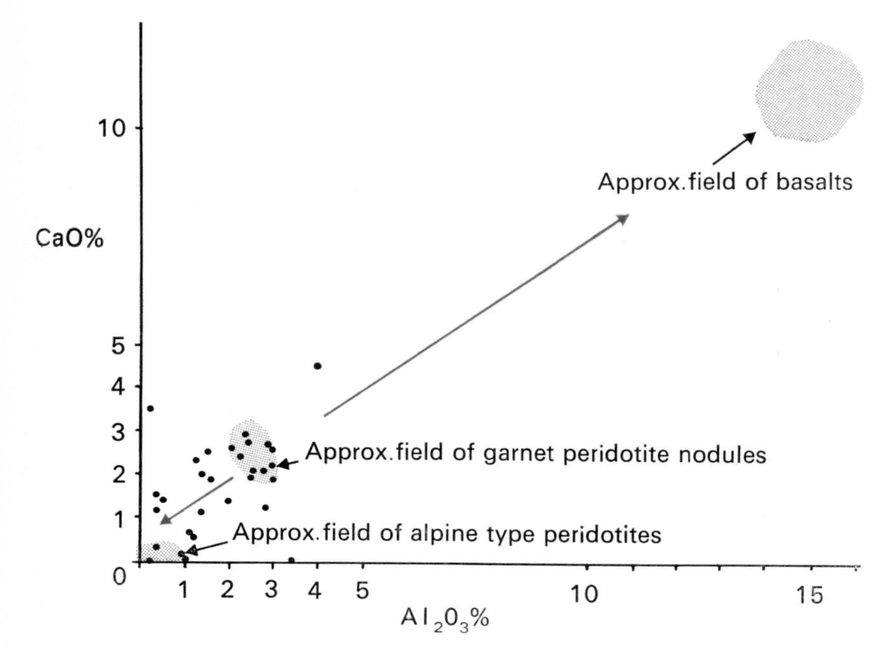

Figure 28 CaO and Al_2O_3 in peridotites of likely mantle origin. The scattered points represent olivine nodules. The approximate fields of Alpine-type peridotites, garnet peridotites and basalts are shown.

Figure 28 plots analysis figures for two of the 'critical' elements in many samples of the three groups of peridotites believed to be of mantle origin. This graph shows the range of chemical variation which the respective averages in Table 10 concealed. As you saw in Unit 1, a range of values is concealed within an average compilation, and in this case the nature of that concealed variation is important.

Applying the principle that, *if a given element is removed from peridotite to form basalt, then the residue will be depleted in that element*, answer the following questions.

SAQ 25

(a) Look at Table 10 and then Figure 28. Which *average* peridotite analysis appears to conceal the greatest range of CaO and Al_2O_3 contents?

(b) Inspect the graph (Fig. 28) and match each of the three peridotite types in the left-hand column with *two* of the descriptions in the right-hand column.

1 Alpine-type peridotites.
2 Olivine nodules.
3 Garnet-peridotite nodules.

A Represent least depleted part of upper mantle.
B Represent most depleted part of upper mantle.
C Represent both depleted and undepleted mantle.
D Least suitable for basalt generation.
E Most suitable for basalt generation.
F Variably suitable for basalt generation.

depleted and undepleted mantle

You have arrived at a plausible (and widely held) interpretation of the range of peridotites which we believe to be samples of the upper mantle. The best samples of 'fresh' mantle, from which basalt has not yet been extracted, are garnet peridotite nodules from kimberlite pipes. On the other hand, Alpine-type peridotites are 'depleted' mantle from which basalt has earlier been removed. The wide range of olivine nodules found in ordinary basalts represents both of these types of mantle.

2.14.4 The average chemical composition of the mantle

In the very remote geological past, before any basalt had been extracted from the mantle – before any continents or ocean crust had formed – the chemical composition of the upper mantle could be considered entirely undepleted, and may therefore have approximated garnet peridotite in composition. Now, however, the upper mantle must contain a mixture of depleted and undepleted peridotite. To estimate its average composition we need to know the balance of these rock types. Since this is something we can never directly determine we must continue to rely upon indirect evidence.

Our basic starting premise is that the analyses we have looked at in this Section are of rocks from the upper mantle. The samples that have come from deepest within the mantle are the garnet peridotites, from depths of about 150 km. The mantle, however, extends to a depth of 2900 km so our samples tell us only about the uppermost part of the upper mantle. We do not know how much of the whole mantle they represent (this depends, among other things, upon how much of the whole mantle is stirred by convection movements, a problem discussed in S100[20]).

In order, therefore, to build a model for the chemical composition of the mantle as a whole, we have to make some assumptions. We assume that the chemical composition of the mantle is similar throughout its great depth range. We also assume that the mass of depleted mantle is negligible when compared with the mass of the mantle as a whole.

By making these assumptions it is possible to use our samples from the upper mantle to make a model composition which can be applied to the mantle as a whole. Although you know that such a composition can be approximated by garnet peridotite nodules, these are rather few in number and an average may not be a very reliable guide to the chemical composition of the mantle. A different procedure has therefore been adopted, one which is built upon a large number of chemical analyses.

It is axiomatic that the mantle must be capable of yielding basalt and leaving a depleted residue upon melting. A recipe for its overall composition can therefore be constructed by using both of these components. Since basalts and Alpine-type peridotites (samples of depleted mantle) are both represented by many analyses and are fairly homogeneous in composition, they can be used for the construction of a mantle model. One such recent model has been constructed by averaging 3 parts of Alpine-type peridotite with 1 part of basalt. Although this choice may seem arbitrary, the composition yielded has appropriate chemical and physical features for the mantle. It is chemically similar to the garnet-peridotite inclusions, but is based upon many analyses of world-wide distribution. Since the mineral composition will, on this chemical basis, consist largely of *pyroxene* and *olivine*, it is known as the *pyrolite* model for the upper mantle.

pyrolite model

Read pp. 35–40 in *Principles of Geochemistry* (The internal structure of the Earth). Note the following points:

1 The structure and composition of the crust are reviewed. Note that we have presented a model for the lower continental crust that is *not* gabbroic (Section 2.13) but its validity remains to be proved.

2 The two hypotheses for the composition of the upper mantle – eclogite and peridotite – are reviewed and the peridotite model preferred.

3 The *chemical* composition of pyrolite is considered to be much the same throughout the mantle. It is the mineral phases, as the physical expression of that composition, which vary from place to place and depend upon changing pressure/temperature conditions. Garnet pyrolite is the type found at deepest levels in the upper mantle and approximates our garnet peridotite. The other mineral assemblages equivalent to pyrolite may be present in some of the wide range of ultramafic nodules which have been found and studied. Positive correlation of a particular nodule with a particular zone in the mantle is, however, difficult and as we noted earlier a certain amount of controversy surrounds the origin of ultramafic nodules.

4 Look at Figure 3.4 in *Principles of Geochemistry* (the distribution of temperatures below oceans and continents).

Is the geothermal gradient steeper below continents or oceans?

Under oceans the geothermal gradient is much greater than below continents – at 300 km below the continents, the upper mantle is inferred to be at a temperature of about 1200 °C, while beneath the oceans it is at about 1500 °C. These marked temperature contrasts in the upper mantle will undoubtedly mean lateral variations in the *mineral* composition of pyrolite, whose *chemical* composition will not change greatly.

5 The detailed nomenclature in the discussion of polymorphic structural types need not concern you (pp. 38–9). You should recognize, however, that as pressures increase some mineral structures change into higher density (i.e. more compact) forms. The changes are of two kinds: (a) purely polymorphic transitions (e.g. olivine ⟶ spinel structures); (b) chemical reactions to form denser phases (e.g. pyroxene ⟶ olivine + stishovite).

mineral reactions in the mantle

6 The density of the mantle does not change regularly with depth. Figure 29 shows one estimate of the increase of density with depth in the Earth's mantle. On the basis of this data the mantle can be divided into three zones.

SAQ 26 Match each item in the first list with one of those in the second:

1 Upper mantle zone.
2 Transition zone.
3 Lower mantle zone.

A Relatively low rate of increase in density.
B Relatively high rate of increase in density.

You should recognize that these changes in density can be explained in terms of the mineral changes inferred to occur within the mantle (cf. note 5 above).

7 The final part of this passage provides an introduction to the next Section.

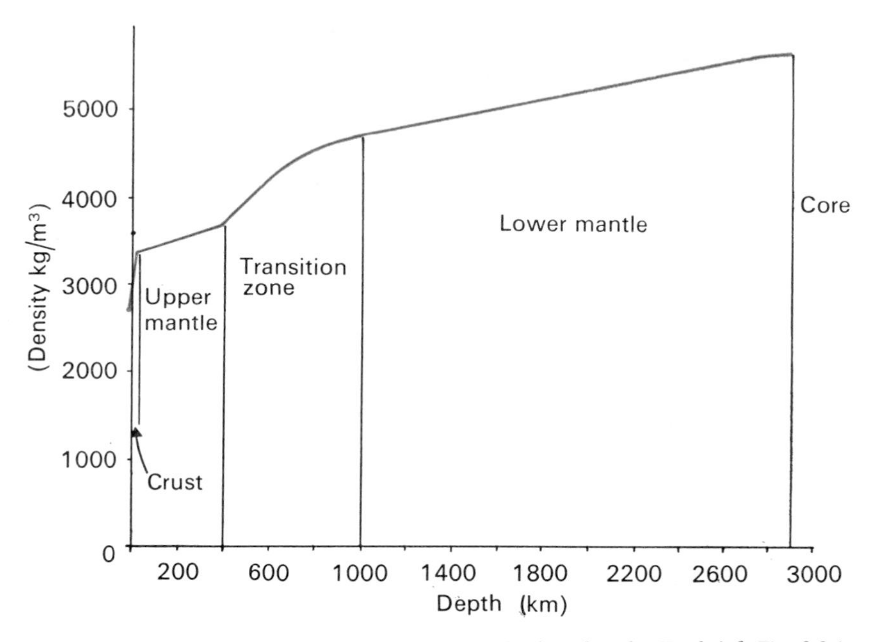

Figure 29 The change in density with increasing depth within the Earth (cf. Fig. 3.2 in Principles of Geochemistry).

SAQ 27 Match each item from the first list with a description from the second:

1 The upper continental crust.
2 The oceanic crust.
3 The core.
4 The lower continental crust.
5 The mantle.
6 Alpine-type peridotites.
7 Ultramafic nodules.

A Small pieces of ultramafic rock, brought to the surface in basalts and kimberlites. They may represent samples of both undepleted and depleted mantle.
B Composed largely of iron or nickel-iron.
C Composed largely of Mg-Fe silicates.
D Olivine-rich intrusions which may represent depleted samples of the mantle.
E Has a bulk composition approaching that of granite.
F Similar in overall composition to continental basalts except for low abundances of potassium and many trace elements.
G May have a composition intermediate between basalt and granite, except for lower abundances of potassium and trace elements.

2.15 The Core *(Objectives 1, 15)*

We know only the very broadest physical features of the Earth's core from seismological studies (S100[21]). Moreover, although we probably have some samples of upper mantle, we are never likely to have any samples of the core, and our knowledge of the composition of the core is based upon our fit of chemical data to this physical framework. Such a fit is, however, based upon extrapolations of experiments performed at lower pressures. For these reasons our knowledge of the chemical features of the core is very speculative.

You were told in S100[21] that the core is composed chiefly of iron, with about 6 per cent nickel, and indeed this was one of the earliest hypotheses, being proposed after the recognition of iron meteorites in the mid-nineteenth century. It is also the most widely accepted current hypothesis but, before looking more closely at it, let us critically examine some other recent models.

An early suggestion was that the core was composed of a condensed form of hydrogen. This was based upon the much greater abundance of hydrogen in the Sun in comparison with the Earth, and the belief that the Earth could retain this hydrogen from the time of formation. You know, however, from consideration of escape velocities, that the Earth could not have retained hydrogen (S100[22]). In addition, the pressure inside the Earth is probably not high enough to cause condensation of hydrogen, so this hypothesis has been discarded.

It was later suggested that the mantle could represent a high pressure modification of mantle material (pp. 39–40 in *Principles of Geochemistry*). This is not popular at the present moment for cogent reasons. Although we cannot carry out static, controlled experiments at the pressures of the core, we can reach very high pressures momentarily, by using explosive-generated shock waves. Even these do not generally reach those of the outermost core, but one experiment achieved a pressure of 2400 kilobars, equivalent to a depth of nearly 4000 km, well within the core.

The results of these experiments carried out on pure minerals and elements can be extrapolated to conditions within the core and used to estimate the density and seismic wave velocities there. From comparison of these estimated properties with the same properties determined by geophysical methods (S100[23]) it is possible to suggest a plausible chemical composition for the core.

The data from these shock wave experiments suggest that silicates would not be dense enough to constitute the core and provide no support for the idea that the core is a high-pressure modification of the mantle. The phases closest in predicted density to that assumed for the core are iron and nickel, a conclusion consistent with the earlier hypothesis (deduced from meteoritic evidence) that the core is composed mainly of iron with a subsidiary content (about 6 per cent) of nickel.

Fe/Ni/light element model for core

There remains, however, an anomaly – the calculated density of the core, although closest to iron containing about 6 per cent Ni, is 10 per cent less than that expected for these metals. The most likely explanation is that the core is largely composed of iron and nickel but contains impurities of low density alloyed with them.

Such impurities must be reasonably abundant and miscible with liquid iron: possibilities include Al, Mg, Si, Na, S, C, O, N, He and H. We can reject He, H, C, O and N, because these elements would not significantly lower the density of iron since they would merely occupy holes already present between the iron atoms. Of the remaining elements, Al, Mg and Na are silicate-forming elements and are more likely to occur in the mantle and crust than the core. This leaves Si and S as distinct possibilities. The choice of Si is more popular than S at the present time but Earth scientists are divided between these two alternatives. In Unit 3 we shall consider evidence which favours sulphur.

2.16 Summary of Sections 2.11–2.15

The geochemical description of the major divisions of the Earth, begun in Unit 1, is now complete.

In Unit 1, evidence was presented that the upper continental crust has a chemical composition approximating that of the common intrusive acid igneous rock type, granodiorite.

The oceanic crust contains three well-defined layers. The surface layer (layer 1) consists of sediments and layer 2 largely of basalts. The nature of layer 3 is not certain but it may contain intrusive and metamorphic rocks of basaltic composition and minor serpentinite masses. Basalt samples collected from the ocean floors are similar in chemical composition to continental basalts but show some important differences – they have very much lower contents of K and some trace elements, notably Rb, Ba, Sr. These chemical features are so distinct that the basalts forming most of the ocean floor are distinguished as oceanic tholeiites.

The nature of the lower continental crust is less certain. Geophysical, geological and experimental data have not unequivocally distinguished between three alternative models:

(a) a dry basic chemical composition (gabbroic);
(b) a wet basic chemical composition (amphibolitic);
(c) a dry intermediate composition (granulitic).

At the present time (c) appears to be the most popular. The postulated lower crust granulite is envisaged as being slightly less acid than the granodiorite upper crust and in having lower contents of K and much lower contents of some trace elements, especially Th and U.

The mantle is the largest division of the Earth. The upper mantle is the source of the basalt magmas erupted on to the sea floor and the continents. These basalts bring up small fragments believed to be of upper mantle material—olivine-rich (peridotite) nodules. Other samples of probable mantle material are Alpine-type peridotites emplaced in belts of mountain building. The third type of mantle sample is the garnet peridotite inclusions brought up in kimberlite pipes. These have come from the greatest depth within the upper mantle. From a study of chemical analyses of these three types of mantle sample it is concluded that the garnet peridotite inclusions represent mantle from which the basalt has *not* been extracted. Alpine-type peridotites appear to represent mantle from which basalt *has* been extracted. Olivine nodules include both types of mantle material.

Since the mantle must be capable of melting to give basalt, leaving a refractory residue, and since we have many analyses of basalt and residual mantle (Alpine-type peridotite), a model for the chemical composition of the upper mantle can be made by assuming it to be composed of 3 parts Alpine-type peridotite + 1 part basalt. This is known as the pyrolite model. Application of this model to the whole mantle can be used to explain the varying distribution of density therein.

The core is completely inaccessible to us. The occurrence of nickel-iron meteorites and inferred physical features of the core suggest that it is composed largely of iron with some alloyed nickel. The density of such a mixture by itself is greater than that deduced for the core, and it seems that a lighter element must be included with this alloy. At present, the most popular choices for such an element are S and Si.

We have now described the major chemical features of the solid Earth. As an additional summary, you should read pp. 40–1 in *Principles of Geochemistry* (The zonal structure of the Earth). This will lead into Unit 3 where we shall consider some of the *processes* which have led to the present distribution of elements throughout the Earth.

Self-assessment Answers and Comments

SAQ 1 Silicon has 4 electrons in its outermost (incompletely filled) shell. Loss of these would give an Si^{4+} ion with the atomic structure of Ne. Aluminium has 3 electrons in the outermost shell—loss of these would give an Al^{3+} ion, again with the atomic structure of Ne. The formation of the O^{2-} ion from oxygen, however, involves the addition of two electrons—these complete the outermost shell and the resultant ion again has the electronic structure of Ne.

SAQ 2

(a) Radius of Si^{4+} is 0.42Å and that of O^{2-} is 1.40Å. The *radius ratio* is therefore 0.42:1.40 or *0.30 to 1*.

(b) The relationship between radius and volume is given by $v=\frac{4}{3}\pi r^3$. The volume is thus proportional to the cube of the radius and the ratio for the volume of Si^{4+} to the volume of O^{2-} is $(0.30)^3$ to 1^3=*0.027 to 1*. You can see from these calculations why oxygen is so abundant by volume in nature. Although its radius is about 3.3 times, its volume is about 37 times that of Si^{4+}.

SAQ 3 The ionic radius of Al^{3+} is 0.51Å at the upper limit of size for 4-coordination and the lower limit for 6-coordination (Table 4.2 in *Principles of Geochemistry*).

SAQ 4 Fluorine (F^-) has a radius of 1.36Å. Since this differs from (OH^-, r=1.40Å) by less than 15 per cent it can substitute for OH^- in the chemical structure of teeth. Chloride, however, (Cl^-, r=1.81Å) has a radius which is over 15 per cent greater than that of (OH^-), and so it cannot be incorporated so easily into the chemical compounds present in teeth.

SAQ 5 KCl and KBr form a complete range of mixed crystals because Cl^- (r=1.81Å) is very similar in ionic radius to Br^- (r=1.95Å). Cl and Br can thus replace each other easily in these alkali halides. For NaCl and KCl, replacement of Na^+ (r=0.97Å) by K^+ (r=1.33Å) is more difficult because these ions have a much greater difference in radius. Hence there is only limited solid solution between KCl and NaCl.

SAQ 6 The alkali feldspars, like the alkali chlorides, will show only partial solid solution between the Na and K end members.

SAQ 7

(a) The olivines form a complete solid solution between Fe_2SiO_4 (fayalite) and Mg_2SiO_4 (forsterite) because the ionic radii of Fe^{2+} (r=0.74Å) and Mg^{2+} (r=0.66Å) differ by less than 15 per cent (see also the last *SAQ* in Unit 1).

(b) There is complete solid solution between $FeCO_3$ and $MgCO_3$. Ca^{2+} (r=0.99A) has an ionic radius which is over 15 per cent greater than either Fe^{2+} or Mg^{2+} and so solid solution between $CaCO_3$ and either $FeCO_3$ and $MgCO_3$ is not complete.

(c) We cannot predict from the information available. As regards ionic radii of cations, Na^+ and Ca^{2+} are so close that solid solution is easily possible. If $(NO_3)^-$ and $(CO_3)^{2-}$ are similar in size, then solid solution should be possible. Natural solid solutions are not known, however, so it is very unlikely that the two are the same.

SAQ 8 Substitution between $2Al^{3+}$ and $3Mg^{2+}$ would be necessary (see also the Section on the micas, p. 21). The radii of these two elements differ by more than 15 per cent – complete solution between these two micas is *not* therefore possible, and is very limited in natural micas.

SAQ 9 Isomorphism is the phenomenon of substances with *different* (but analogous) chemical formulae having similar crystal structures (they are isostructural). Polymorphism, on the other hand, refers to the occurrence of a mineral with fixed chemical composition in *different* crystal forms.

SAQ 10 Because it changes back into graphite at an infinitesimal rate under ordinary atmospheric pressure and temperature conditions (p. 92 in *Principles of Geochemistry*).

SAQ 11

(a) Your stability graph should qualitatively resemble Figure 30.

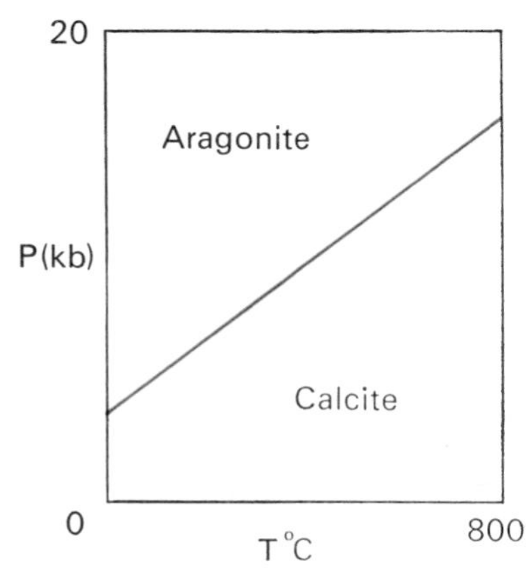

Figure 30 The stability fields of calcite and aragonite.

(b) Aragonite is stable at higher pressures and therefore it is likely to be (and is) the denser polymorph of $CaCO_3$.

(c) Inversion to calcite is comparatively slow – but not so slow as that of diamond to graphite – hence the aragonite is preserved. But the inversion rate is such that no aragonite is found in rocks older than about 100 Ma.

SAQ 12

(a) There are 3 silicate tetrahedra, which would make a total of Si_3O_{12} with net charge of 12–. BUT 2 oxygens are shared (the black ones in your model), so we have a formula of Si_3O_{10}.

(b) Those two black oxygens are neutral, moreover, leaving 8 charged oxygens (the red ones), so net charge is 8–. You can also work this out by assigning 4+ to Si and 2– to O. Thus, $3\times4=12+$, and $10\times2=20-$, giving a net charge of 8–.

SAQ 13 If each oxygen atom is shared between *two* tetrahedra the number of effective oxygens per silicon atom is halved. Moreover, an oxygen shared between two silicons is always neutral. In quartz, every oxygen is shared in this way. Therefore, with the number of effective oxygens halved, and all of these oxygens neutral, the overall formula becomes SiO_2. ($2\times Si^{4+}$= 8+; $4\times O^{2-}$=8–. This gives a net charge of nil.)

SAQ 14 If one in four silicons (Si^{4+}) in quartz is replaced by aluminium (Al^{3+}) then there will be a charge deficiency of one positive charge per four anionic groups. This can be satisfied by the incorporation of one monopositive cation into the structure. The most abundant such ions are Na^+ and K^+ and the addition of these ions to the aluminosilicate anionic structure results in the formation of the alkali feldspar group.

SAQ 15 Feldspar, pyroxene and olivine (in that order) are the major constituents of basalt. Feldspar, quartz and mica (in that order) are the major constituents of granite.

SAQ 16
(a) 1B, 2A, 3C. (b) 1B, 2C, 3A.

SAQ 17
1A Iceland lies athwart the mid-Atlantic ridge.
2C The Canary Islands lie just off the coast of N. West Africa, which is part of a large composite oceanic and continental plate. The nearest plate margin to the Canaries is the mid-Atlantic ridge.
3B Tristan lies close to the mid-Atlantic ridge.
4D The Lesser Antilles lie close to a destructive plate margin. The chain of islands can be traced around into the north-eastern part of the Andes.

SAQ 18
1 Layer 1 (b) (iii)
Layer 2 (a) (ii)
Layer 3 (c) (i)
2 Oceanic tholeiite basalt has significantly lower concentrations of K_2O (major element) and Rb, Ba and Sr (trace elements) in comparison with continental basalts.

SAQ 19 Assume 100 g of gneiss is under consideration. This contains 2 per cent $K_2O=2$ g K_2O. One-quarter of the gneiss is melted to form 25 g of granite with a 4 per cent K_2O content$=$ 1g K_2O.
Thus, as grammes K_2O in gneiss$=$grammes K_2O in granite$+$ grammes K_2O in residual granulite,
or grammes K_2O in residual granulite$=$
grammes K_2O in gneiss$-$grammes K_2O in granite,
then grammes K_2O in residual granulite$=2-1=1$ g.
But weight of residual granulite$=100-25=75$ g.

$$\text{Therefore percentage } K_2O \text{ in residual granulite} = \frac{1}{75}\times 100 = \frac{4}{3}$$
$$= 1.33 \text{ per cent.}$$

SAQ 20
(a) $2MgO.SiO_2$
(b) Atomic weight Mg$=24.32$
O $=16.00$
Si $=28.09$
Molecular weight $2MgO=2\times(24.32+16.00)=80.64$
Molecular weight $SiO_2=28.09+(2\times 16.00)=60.09$
Molecular weight $2MgO.SiO_2$ or $Mg_2SiO_4=140.73$

$$\text{Weight per cent } MgO = \frac{80.64}{140.73}\times 100 = 57.3 \text{ per cent}$$

$$\text{Weight per cent } SiO_2 = \frac{60.09}{140.73}\times 100 = 42.7 \text{ per cent}$$

SAQ 21 Both reasons are true. Natural olivine always contains some Fe substituting for Mg. Natural peridotites very rich in olivine are called dunites, but even these always contain small amounts of other minerals besides olivine.

SAQ 22 Alternative 3 is correct. A geothermal gradient of 25 °C/kb would approximate a straight line passing up through 1300 K/40kb to intersect the phase boundary at about 50 kb. This corresponds to a depth of about 150 km, which is well within the upper mantle. In fact, the geothermal gradient is approximately linear down to 100 km, after which its slope decreases. But this does not significantly affect our conclusion.

SAQ 23 You should have ticked TiO_2, Al_2O_3, CaO, Na_2O and K_2O.

These oxides are all very much higher in basalts than in any type of peridotite.

SAQ 24
(a) If all the K_2O in the original Alpine-type peridotite (0.005 per cent) goes into the basalt (0.9 per cent) there has been a concentration factor of 0.9/0.005 or 180. In other words, the K_2O is distributed in a mass of basalt which is 1/180th of the mass of the original peridotite. This is:

$$\frac{1}{180}\times 100 = 0.55 \text{ per cent of the original peridotite.}$$

(b) By similar reasoning, the production of basalt from garnet peridotite requires a concentration factor of 0.9/0.03 or 30. The mass of basalt is 1/30 or 3.33 per cent of the original peridotite.
(c) Six times as much basalt can be obtained from the garnet peridotite as from the Alpine-type peridotite.
(d) Garnet peridotite, of course.

SAQ 25
(a) The analyses of olivine nodules have much more variable contents of CaO and Al_2O_3 than the Alpine-type or garnet peridotite analyses shown. An average of these elements in olivine nodules thus conceals a greater variation than in other peridotite types.
(b) 1, B, D. Alpine-type peridotites have uniformly lower contents of CaO and Al_2O_3 than the other peridotite types. Since these two oxides are enriched in basalts we think that Alpine-type peridotites represent mantle samples from which basalt might have been generated in the past, so that they are now depleted in the chemical components of basalt and are no longer suitable for basalt generation.
2, C, F. Olivine nodules cover a wide range of variation in Figure 28, from high to low contents of CaO and Al_2O_3. Clearly for the low CaO and Al_2O_3 nodules, what we have just said about the Alpine-type peridotites applies. For the nodules which have higher contents of these elements, see below.
3, A, E. Because garnet peridotite nodules have consistently higher contents of CaO and Al_2O_3 than the other peridotite types, they represent mantle samples which are so far undepleted and therefore still suitable for yielding basalt by partial melting.

SAQ 26 1A; 2B; 3A.

SAQ 27 1E; 2F; 3B; 4G; 5C; 6D; 7A.

Appendix 1 References to S100 Units

1 Unit 8, *The Periodic Table and Chemical Bonding*, Section 8.4.9

2 Unit 8, Section 8.5.2; Unit 13, *Giant Molecules*, Section 13.1.1

3 Unit 7, *The Electronic Structure of Atoms*, Section 7.1.4

4 Unit 8, Sections 8.4.1 and 8.4.4

5 Unit 8, Appendix 2

6 Unit 8, Section 8.4.6

7 Unit 8, Section 8.4.1

8 Unit 13, Section 13.1.1

9 Unit 8, Section 8.5.2

10 Unit 5, *The States of Matter*, Home Experiment

11 Unit 10, *Covalent Compounds*, Section 10.5.2

12 Unit 22, *The Earth, Its Shape, Internal Structure and Composition*, Section 22.5

13 Unit 22, Section 22.5.8; Unit 25, *Continental Movement, Sea-floor Spreading and Plate Tectonics*, Section 25.1.4

14 Unit 24, *Major Features of the Earth's Surface*, Atlantic Floor Chart

15 Unit 25, Sections 25.1.4 and 25.2

16 Unit 22, Section 22.5.8

17 Unit 22, Section 22.5.9

18 Unit 22, Section 22.4.6

19 Unit 8, Section 8.5.2 and TV24

20 Unit 25, Section 25.2.4

21 Unit 22, Sections 22.5.4, 22.5.5 and 22.5.6

22 Unit 27, *Earth History II*, Section 27.1.4

23 Unit 22, Section 5.5

Appendix 2 Glossary

ANDESITE A volcanic rock of intermediate chemical composition, containing about 60 per cent SiO_2.

DOLERITE An intrusive igneous rock similar to basalt in composition but slightly coarser-grained.

GABBRO A coarse-grained igneous rock of basic chemical composition which occurs in large intrusions, and is the chemical equivalent of basalt.

GEOTHERMAL GRADIENT The rate of increase of temperature with depth within the Earth.

GNEISS A coarse-grained metamorphic rock which commonly has bands rich in light coloured minerals (quartz and feldspars) alternating with bands rich in dark minerals (e.g. hornblende and biotite).

ISOELECTRONIC SERIES A series of molecules or ions which fall in the same horizontal row in the Periodic Table and contain the same number of electrons, e.g. Na^+, Mg^{2+}, Al^{3+}.

LAW OF CONSTANT PROPORTIONS The law which states that a given chemical compound always has the same composition, no matter how it formed.

RHYOLITE An acid volcanic rock, equivalent to granite in composition.

ULTRAMAFIC ROCK This igneous rock contains a large proportion (over 70 per cent) of mafic minerals (rich in Mg and Fe). Since minerals rich in MgO and FeO are also often low in SiO_2, ultramafic rocks are very usually but not exclusively ultrabasic in chemical composition.

VAN DER WAALS FORCES Weak physical forces of attraction *between* uncharged atoms and molecules which are responsible for their cohesion. These forces are distinct from those of chemical bonding *within* the molecules.

Appendix 3 (Black)

Radius Ratios

If the anions in a coordination polyhedron just touch the cation and each other, we can find, by using simple trigonometry, the radius ratio (of cation to anion) for each shape or coordination number. To show how this is done, we shall calculate the radius ratio (r_c/r_a where r_c=radius of cation and r_a=radius of anion) for tetrahedral coordination. Figure 31 shows a tetrahedron, drawn by connecting the centres of the four anions, A, B, C, D, by straight lines (cf. Fig. 1). The outline of the cation is drawn in and its centre-point O is shown connected by straight lines to A, B, C and D. AOB is the tetrahedral angle, which can be shown to be 109.5°.

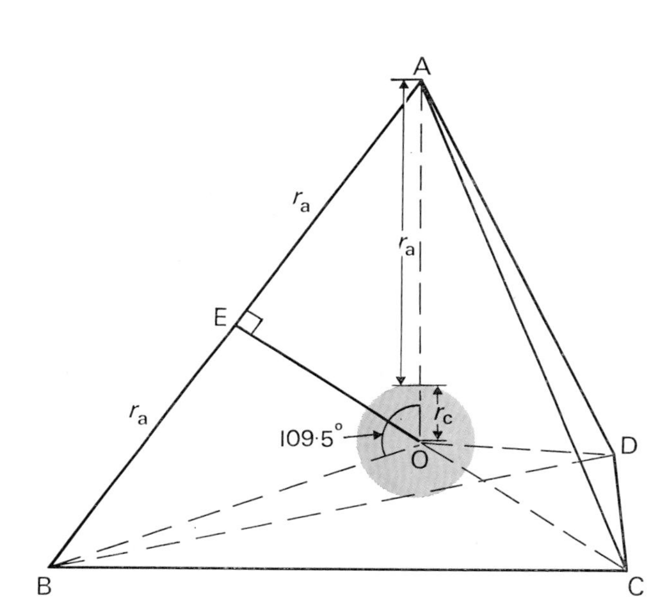

Figure 31 The radius ratio for tetrahedral coordination.

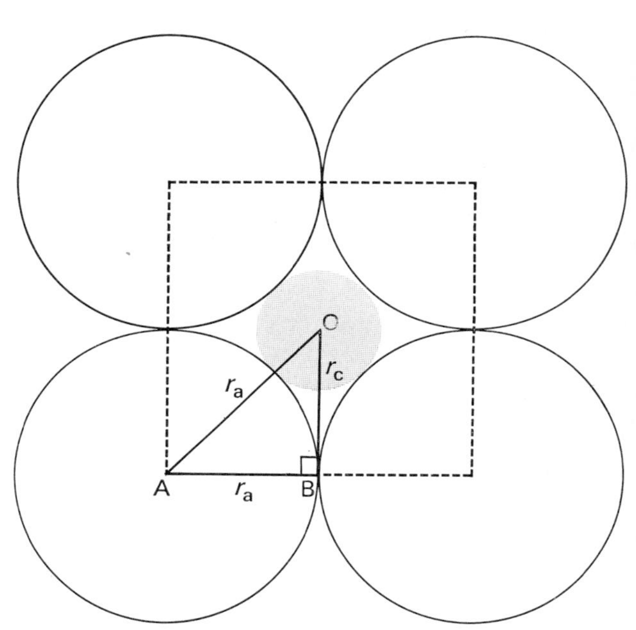

Figure 32 The radius ratio for octahedral coordination.

Now because AB runs between the centres of two anions which are just touching, it is equal to $2r_a$. The centre point of AB is thus marked by E and the line EO is marked. Because AO=BO, then EO is a perpendicular from AB to O.
From a study of the triangle AEO:

$$\frac{AE}{AO} = \sin\left(\frac{109.5}{2}\right)$$

therefore AE = AO sin 54.75°,
but AE = r_a and AO = $r_a + r_c$,
therefore $r_a = (r_a + r_c)$ sin 54.75°.
This gives $r_c = 0.22\ r_a$ which is called the *limiting radius ratio* for tetrahedral coordination. As r_c/r_a increases there comes a point at which 6 anions can cluster around the central cation. A square cross-section of this arrangement is shown in Figure 32.

The limiting radius ratio for octahedral coordination can be calculated from this diagram. This can be done without using trigonometric tables – since AOB is a right-angled triangle you can use Pythagoras' theorem to calculate the relationships between r_a and r_c. If you like, try this calculation now. The explanation runs as follows.

Consider the triangle AOB (Fig. 32) in which OAB = 45°

$$\frac{AB}{AO} = \frac{r_a}{r_a + r_c} = \cos 45°$$

$$\therefore \quad r_a = (r_a + r_c) \cos 45°$$

$$r_a = \cos 45°\, r_a + \cos 45°\, r_c$$

since $\cos 45° = 0.71$

$$r_a = 0.71\, r_a + 0.71\, r_c$$

$$r_a - 0.71\, r_a = 0.71\, r_c$$

Now, dividing through by r_a gives:

$$1 - 0.71 = \frac{0.71\, r_c}{r_a}$$

$$r_c = \frac{r_a\,(1 - 0.71)}{0.71}$$

$$r_c = \frac{0.29}{0.71}\, r_a$$

$$r_c = 0.41\, r_a$$

The radius ratio for when the ions just touch sets a rough minimum for that coordination number. As a cation increases in size, the anions move outwards (so to speak) until there is room for more anions to coordinate, i.e. we move into the range of the next larger coordination polyhedron. But if a cation is smaller in size than the radius ratio of a particular polyhedron demands, so that it rattles about (so to speak) in the cavity, then that polyhedron is obviously unstable relative to the next smaller one. The information in Table 11 has been derived in this way.

Table 11

Coordination number and type	*Radius Ratio* *Lower limit*	*Range*
3 triangular	0.15	0.15–0.22
4 tetrahedral	0.22	0.22–0.41
6 octahedral	0.41	0.41–0.73
8 cubic	0.73	0.73–1
12 close-packing	1	1

(Compare Table 4.1 in *Principles of Geochemistry.*)

In minerals, the most important anion is oxygen, either as O^{2-} in silicate or as OH^- (although other anions such as F^- and S^{2-} are not uncommon). Assuming for the moment that silicate structures are assemblages of ions such as O^{2-}, Si^{4+}, Al^{3+}, B^{3+}, Na^+, K^+, Mg^{2+}, Ca^{2+} and Fe^{2+}, we can determine the radius ratios of the cations relative to the oxide ion (from Table 1 in the main text). This has been determined in Table 12 using radii in octahedral coordination. Do you see any fault in logic in this procedure?

Table 12

Ion	r_c	r_c/r_a	*Coordination number* *expected*	*found*
B^{3+}	0.23	0.16	3	3, 4
Si^{4+}	0.42	0.30	4	4
Al^{3+}	0.51	0.36	4	4, 6
Fe^{3+}	0.64	0.46	6	6
Mg^{2+}	0.66	0.47	6	6
Fe^{2+}	0.74	0.53	6	6
Na^+	0.97	0.69	6	6, 8
Ca^{2+}	0.99	0.71	6	6, 8
K^+	1.33	0.95	8	8, 12

(Compare Table 4.2 in *Principles of Geochemistry.*)

In Table 12 are given, also, the coordination numbers found in minerals. In all cases, the expected coordination number is found, and, in some cases, others are also possible. Thus aluminium is 4-coordinate as aluminate, but 6-coordinate when it replaces Mg^{2+}, etc., in micas. Boron is 3- and 4-coordinated by oxygen in the same structure, in borax for example. A fault in our procedure is that the ionic radii in Table 1 of the main text are those for 6-coordination, and have been used in some cases to predict 3-, 4- and 8–12-coordination. Strictly, this is invalid, since the radii of our supposed ions vary with coordination. Thus the Al–O distance in the aluminosilicates is 1.77 Å for 4-coordination (aluminate) and 1.8–2.1 Å for 6-coordination (Al^{3+}).

Ionic radii are usually quoted for 6-coordination, since they then form a set consistent with many 'standard' structures, such as those of alkali halides. It is much more difficult to apportion cationic and anionic radii in the tetrahedral or triangular clusters, since a cation which is small and (formally) highly charged, distorts or polarizes neighbouring anions by strongly attracting their electrons. The bonding is then relatively covalent, and the 'ionic model', of charged incompressible spheres, begins to break down.

For example, the Si–O bond length in the silicates varies remarkably little from the olivines to the feldspars, and is usually about 1.62 ± 0.02 Å. But this is much less than the sum of the Si^{4+} and O^{2-} radii in Table 1, namely 1.82, and this is due to the tetrahedral coordination, and to the covalent character. The length, and therefore the bond type, is clearly more variable for the Al–O bond.

All the same, the ionic model is a very useful one in this context, as long as we do not expect more than semi-quantitative results from it.

Appendix 4 (Black)

Pauling's Rules for Ionic Structures

These rules (which include principles developed by Goldschmidt and by Bragg) were induced from crystal structures and deduced from the equations that express the energy, and therefore the stability of an ionic crystal, in terms of the radii, charges, and arrangement of the ions. These rules are not rigorously obeyed, but point to the most stable arrangements; they have been very useful in the working out of mineral structures by X-ray crystallography, and have helped our understanding of their unexpected properties. These rules just state, in a rather more formal way, many of the principles which you have used in studying Sections 2.1–2.10 of this Unit. Rule 1 is already familiar.

1 Around each cation, the anions form a *coordination polyhedron* in such a way that the cation-anion distance (centre to centre) is the sum of the ionic radii; the coordination number is determined by the radius ratio.

2 *The Electrostatic Valence Rule.* A cation M^{x+}, coordinated by y anions, contributes x^{+}/y charge to neutralize each anion. Each anion A^{z-} receives a total charge z^{+} from the cations that are adjacent (i.e. at the centres of all the polyhedra of which this anion forms a part).

This criterion ensures that charges are neutralized in as small a space as possible, to maximize attractive forces and minimize repulsions. According to Pauling, this rule is followed by most of the known silicate structures, and deviations by as much as $\pm 1/6$ of an electronic charge are rare. The larger deviations are found in synthetic compounds which are not as stable as minerals.

This rule has many applications, for example to estimate the relative strengths of ionic bonds. The simple arithmetic of this rule greatly limits the number of acceptable structures for a given composition, and sometimes specifies the structure uniquely. In the feldspars, for example, it explains that large cations with rather high coordination numbers (such as Na^{+}, Ca^{2+}, and K^{+}) can be accommodated, but not small highly-charged ions such as Fe^{2+} or M^{2+}, because of the low charge on the anion framework.

This rule also explains the stability of the giant polyanion silicates, alumino-silicates, and borates, in which the shared oxygens are neutral or slightly negatively charged. Sulphur and phosphorus polyacids and polyanions do exist in which the tetrahedra share oxygen, but there are no sulphate or phosphate minerals of this type. This is because of the ready hydrolysis to orthophosphates or sulphates, in which the electrostatic valence rule is more readily satisfied.

3 A structure is less stable if anion polyhedra share edges, and less stable still if they share faces. This effect is greater the smaller the cation, and the larger its charge.

Model exercise H

You can study this rule semi-quantitatively by using the Home Kit ball-and-spoke models. For this you will need to make three plasticine balls of similar size to the plastic balls. In this experiment you must measure the cation–cation distance in three different silicate polyanions:

(a) With one vertex shared, as in $Si_2O_7^{6-}$ (cf. Figs. 5 and 8). Assume that the Si–O–Si bond is 180° and use a plasticine ball for the linking oxygen.
(b) With two vertices shared – i.e. with one edge in common. For this you will have to use two plasticine balls for the linking oxygens.
(c) With three vertices shared – i.e. one face in common. For this, use three plasticine balls for linking oxygen. After you have measured the cation–cation distances as carefully as practical in these three cases, calculate the ratios of these distances in (b) and (c) to the distance in (a). What (rough) estimate does this give of the decrease in stability in passing from (a) to (b) to (c)? How does this depend (algebraically) upon the size and charge of the cation?

The ratios are 1:0.58:0.33. The decrease in stability is mainly due to the increase in cation–cation repulsion (although there is also a decrease in the extent to which the anions shield the cations from each other). On this basis the cation repulsion energies (in the isolated double-tetrahedra) are given by z^2e^2/r, where z is the valency of the cation and e the electronic charge. The potential energy thus increases as the reciprocal of the cationic separation, r, i.e. as 1:1.7:3, with consequent *de*crease in stability.

The potential energy due to cation repulsion increases as the square of the charge, and increases also as the cation gets smaller, as this reduces r.

It is interesting to note that, for octahedral coordination, these ratios are 1:0.71:0.58, i.e. the decrease in stability is not so pronounced when the cations are (relatively) further apart.

The silicates exemplify this third rule. Silicon as a cation is small and highly charged, with the result that the silicate tetrahedra can share vertices, but not edges or faces. When structures appear to break this rule, it turns out that the bonding is covalent rather than ionic, so that Coulomb repulsion is small anyway. An example is the sulphur 'analogue' of silica, SiS_2, in which the tetrahedra share edges to form infinite chains:

```
      S      S      S      S
     / \    / \    / \    / \
... Si   Si    Si    Si    Si ...
     \ /    \ /    \ /    \ /
      S      S      S      S
```

silicon sulphide

But the electronegativities of silicon (1.8) and sulphur (2.5) are not very different, and the atoms cannot be considered to be ions.

4 Cations that are small and highly charged tend not to share polyhedron elements with each other. This is related to the earlier rules, and to the discussion of polyphosphates, etc. Such atoms tend to form bonds that are covalent rather than ionic. Ionic structure types are usually determined by the anions, since these are bigger than the cations. This is particularly true of the silicates, which consequently have a remarkable variety of polyanions.

5 *The Principle of Parsimony.* The number of different kinds of atomic site in a crystal tends to be small.

A

B

C

D

E

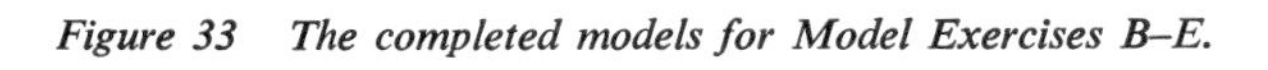

Figure 33 The completed models for Model Exercises B–E.

F

G

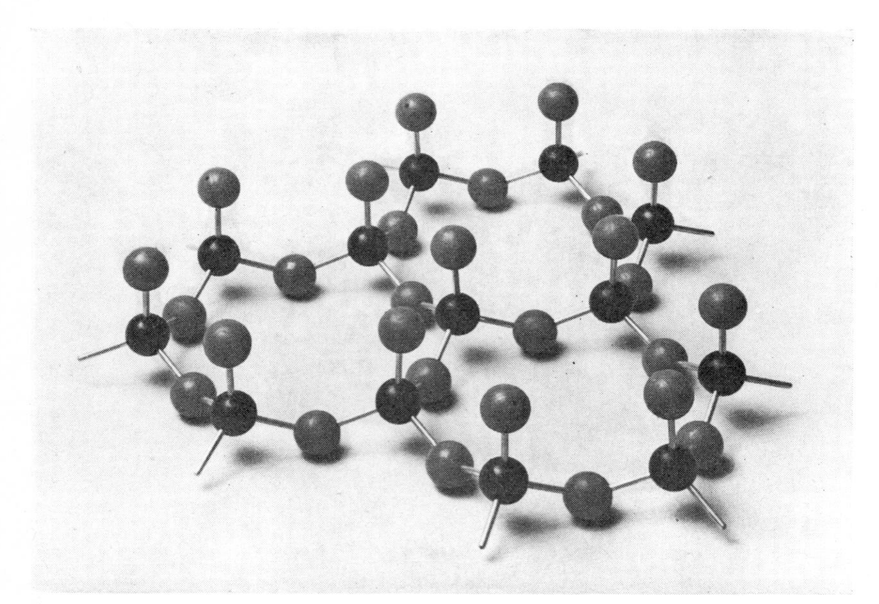

H

Figure 33 The completed models for Model Exercises E–G.

Acknowledgements

Grateful acknowledgement is made for illustrations used in this Unit:

Figure 7: George Allen & Unwin, from F. H. Hatch, A. K. Wells and M. K. Wells, *Petrology of Igneous Rocks*; Figure 10: Cambridge University Press, from R. C. Evans, *An Introduction to Crystal Chemistry*, 2nd edition, 1964; Figures 12a and 14a: G. Bell and Sons Ltd, from C. Bragg and G. F. Claringbull, *Crystal Structures of Minerals*; Figure 19: Charles E. Merrill Publishing Co, from R. H. Foster, *Geology*; Figure 20: Acta Crystallographica for E. J. W. Whittaker in *Acta Crystallographica*, **10**, 155, 1957; Figure 21: Macmillan Journals Ltd, from J. R. Cann, 'New model of the structure of the ocean crust' in *Nature, Lond.*, **226** (5249); Figure 22: Geological Survey of Canada; Figure 24: W. H. Freeman & Company, from H. Williams, F. J. Turner and C. M. Gilbert, *Petrography*; Figure 25: Elsevier Publishing Company, from I. B. Lambert and K. S. Heier, 'Estimates of the crustal abundances of thorium, uranium and potassium', in *Chemical Geology*, **3**, 1968; Figure 28: American Geophysical Union, P. G. Harris, A. Reay and I. G. White, 'Chemical composition of upper mantle' in *Journal of Geophysical Research*, **72** (24).

UNIT 3
THE GEOCHEMICAL HISTORY OF THE EARTH

Contents

Table A

List of Scientific Terms, Concepts and Principles used in Unit 3

Taken as prerequisites				Introduced in this Unit			
Assumed from general knowledge (GK) or from S100*	**Unit No. or GK**	**Defined in a previous second level Unit**	**Unit No.**	**Defined and developed in this Unit**	**Page No.**	**Introduced here, but fully developed in later Unit**	**Unit No.**
absorption line	6	granodiorite chemical composition of upper continental crust	1	geochemical classification	8	partial melting	4
achondrite	27	absorption and emission spectrographic analysis	1	siderophile	8	origin and development of the continental crust	5
atomic number	6	geochemical standards	1	chalcophile	8	growth of atmosphere and hydrosphere	5
asteroid	22	granulite model for lower continental crust	2	lithophile	8		
basalt	22	oceanic tholeiite model for oceanic crust	2	atmophile	8		
biosphere	20	peridotite mantle model	2	chemical composition of the whole Earth	12		
carbonaceous chondrite	27	metallic core model	2	chemical composition of the Sun	14		
chondrite	27	olivine	2	photosphere	15		
covalent bond	8	pyroxene	2	model photosphere	15		
granite	26	serpentine	2	cosmic element abundances	16		
electron	6	albite-anorthite plagioclase	2	classification of meteorites	16		
electron volt (ev)	6	garnet	2	Widmanstätten structure	16		
electron shell	6, 7	eclogite	2	meteorite finds and falls	16		
exothermic	11, 12	peridotite inclusions in basalt	2	chondrites	18		
exponent	2, MAFS**			chondrules	18		
heat of formation	11, 12			chondrite groups	20		
immiscibility	GK			ordinary chondrites	20		
ionic (electrovalent) bond	8			C1 carbonaceous chondrites	20		
ionic radius	8			cosmic abundance pattern	20		
ionization	6			Oddo-Harkins rule	20		
ionization potential	6, 7			significance of chondrules	22		
iron (meteorite)	27			trace element fractionation in chondrites	22		
island arc	24						
isotope	26						
lunar Mare	27						
meteorite	27						
Moho	22						
neutron	6						
nuclear reaction	6, 31						
nuclide	6						
oxidation	8						
Periodic Table	8						
photosynthesis	15						
planet	GK						

* *The Open University (1971) S100* Science: A Foundation Course, *The Open University Press.*

** *The Open University (1970) S100:* Mathematics for the Foundation Course in Science, *The Open University Press.*

Taken as prerequisites				Introduced in this Unit			
Assumed from general knowledge (GK) or from S100*	**Unit No. or GK**	**Defined in a previous second level Unit**	**Unit No.**	**Defined and developed in this Unit**	**Page No.**	**Introduced here, but fully developed in later Unit**	**Unit No.**
principle quantum number	6			chondritic and solar abundance patterns	22		
proton	6			features of C1 chondrites	23		
radioactivity	6			chondritic Earth model	24		
reduction	8			lunar regolith	25		
seismic P wave	22			lunar anorthosite	25		
seismic S wave	22			radioactive heating of Earth	28		
seismic discontinuity	22			formation of core	29		
Solar System	22			two-stage origin for Earth	29		
spectroscope	6			single-stage origin for Earth	29		
stony-iron meteorite	27			oxidation state of chondrites	30		
transition element	8			carbonaceous reducing agents	30		
valence	8			chemical partition of elements between core and mantle	32		
wavelength	2			relative ease of oxidation of elements	32		
				primary geochemical differentiation	34		
				secondary geochemical differentiation	34		

Objectives

When you have completed this Unit you should be able to:

1 Define in your own words, or recognize valid definitions of the terms, concepts and principles listed in column 3 of Table A.

2 Given chemical data from the Earth, the meteorites and dynamic Earth analogues such as metal smelters, explain how a geochemical classification of the chemical elements can be made into siderophile, chalcophile, lithophile and atmophile groups.

3 Show how the geochemical classification of the elements is related to the fundamental chemical properties of the elements.

4 Outline the difficulties involved in the estimation of cosmic abundances from studies of the Sun's spectrum.

5 Give reasons for the importance of the chondritic meteorites, and outline the criteria used in selecting the meteorite group most closely representative of the non-volatile matter of the solar system.

6 Make, or recognize, statements concerning the chemical features of the Sun, the chondritic meteorites and the Earth as a whole.

7 Explain how samples of rock from the Moon's surface differ from other extra-terrestrial chemical samples and suggest how such samples may differ from the Moon as a whole.

8 Distinguish between the primary and secondary geochemical differentiation of the Earth. List the major stages in these differentiation processes.

9 Outline the physico-chemical factors involved in the separation of a homogeneous Earth into metal, sulphide and silicate portions.

10 Given data concerning the physical and chemical properties of hypothetical and real planets and satellites, make predictions or deductions about the distribution of chemical elements therein.

Conceptual Diagram

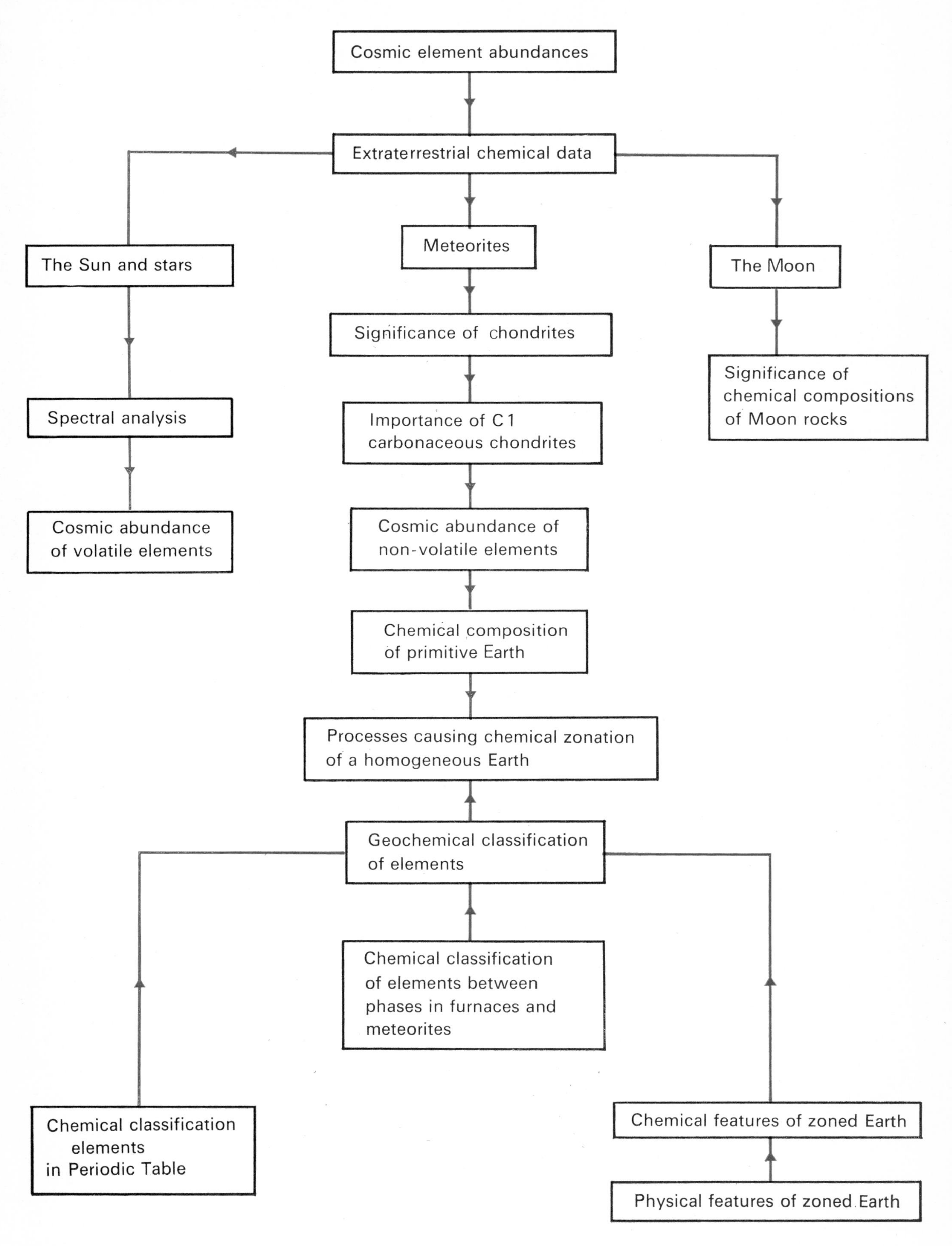

3.1 Introduction

We have so far been concerned with the collection of geochemical data and their limitations, and the use of such data to build up a chemical picture of the Earth and its constituent minerals. At this point we ask: what processes have caused the chemical variations in and upon the Earth? These processes and the geochemical reactions controlling them form the subject of both this Unit and the next.

In this Unit you will look at the ways in which the chemical elements are distributed throughout the Earth and then at the processes by which the Earth might have evolved from its parent materials into its present strongly zoned state. Let us begin on the largest scale. In S100[1]* you studied the physical features of the Earth.

Can you remember what pieces of evidence suggest that the Earth is not uniform in its physical features?

The average density of the surface rocks (2 900 kg m^{-3}) and of the upper mantle (3 300 kg m^{-3}) is much less than that of the whole Earth (5 500 kg m^{-3}). Also, the seismic P and S wave velocities are greater than would be predicted on the basis of a homogeneous Earth. It is possible to use seismic data to build up a fairly detailed picture of the Earth, characterized by discontinuities between shells with different physical and chemical properties. Although there are several discontinuities (and more are being discovered) the most important of these are between the crust, the mantle and the core. The chemical significance of these divisions has been described in the first two Units of this Course. The continental crust has a chemical composition intermediate between granite and basalt. The oceanic crust has the chemical composition of basalt (oceanic tholeiite) and the mantle is composed of an ultrabasic rock – a variety of peridotite. The core, on the other hand, appears to be metallic and is believed to be composed chiefly of iron with nickel and some other elements in small amounts.

Has the Earth always had this layered structure? We have suggested that the Earth formed by accretion of cold planetary material (S100[2]). This suggests two possible origins for the present structure of the Earth:

1 The first particles to accrete were metallic and formed the core, later particles were of silicates and formed the mantle and crust. In this model, accreting iron particles would stick together because they are plastic over a wide range of temperature. Silicates, however, are generally brittle and would not accrete until a large metallic body had been established. The first silicate particles would then become embedded in the metal and accretion would continue owing to the gravitational attraction of the planet.

2 All the accreting particles had the same composition and formed an initially homogeneous Earth, within which development of the core, mantle and crust took place by later processes.

The second hypothesis is popular at the present time but, before looking at it in detail, let us first examine the ways in which elements are distributed between different phases in heterogeneous systems such as the Earth.

**The Open University (1971)* S100 Science: A Foundation Course. *The Open University Press. Superscript numbers indicate more detailed references to S100, which are listed in Appendix 1.*

3.2 The Geochemical Classification of the Elements

(*Objectives 1, 2*)

Read pp. 55 – 9 in *Principles of Geochemistry** (The geochemical classification of the elements). Note that the Periodic Table (Table 3.11) is similar to the one in S100[3]. For the second (right-hand) Y, read Zr.

geochemical classification

Glossary for this reading

ATOMIC VOLUME	The volume of an atom calculated from its atomic radius.
DISTRIBUTION COEFFICIENT	In a chemical system containing solid and liquid phases (e.g. a crystallizing lava), the ratio of the equilibrium concentration of an element in the solid to its concentration in the liquid phase is known as the *distribution coefficient*.
ELECTRODE POTENTIAL	(Defined in Section 3.3).
MANSFELD COPPER SLATE	A copper ore from Mansfeld in Germany.
MATTE	An immiscible sulphide-rich phase, which separates out during the smelting of certain ores.
NEUTRON ACTIVATION ANALYSIS	A highly accurate method of element analysis. The amount of an element present in a sample is determined by first irradiating the sample with suitable nuclear particles, then measuring the intensity of characteristic radioactivity induced in the element. The intensity of this induced radioactivity is directly proportional to the amount of the element present.
TROILITE	Meteoritic iron sulphide (FeS). The commonest terrestrial iron sulphide, pyrite, is FeS_2.

The important points to note in this Section are:

1 The geochemical classification of the elements is based on their affinity for a particular primary phase. The terms used are listed below:

Siderophile – affinity for metallic iron
Chalcophile – affinity for sulphide
Lithophile – affinity for silicate
Atmophile – affinity for the gaseous state (at surface P,T conditions)

siderophile
chalcophile
lithophile
atmophile

2 The geochemical classification can be applied to the distribution of elements between metal, silicate and sulphide phases in mixed systems, such as the Earth, meteorites and smelter products. The last two systems are accessible and provide data for a quantitative classification which may be applied to the Earth as a whole. Table 1 shows the abundance of selected elements in different phases of meteorites.

SAQ 1 Elements often show a variable tendency to enter more than one phase. Match the numbered and lettered columns by using the data from Table 1.

1 Cu
2 Mg
3 Au
4 Th
5 Pb
6 Os
7 Ca
8 Pt

A strongly lithophile
B dominantly chalcophile, but shows siderophile or lithophile tendency
C dominantly siderophile

* *Brian Mason (1966)* The Principles of Geochemistry, *3rd ed. John Wiley (paperback).*

Table 1 Distribution of some elements between the metal, sulphide and silicate phases of meteorites. Concentrations are in parts per million

	metal (nickel-iron)	*sulphide (troilite)*	*silicate*
Ca	—	—	16 700
Th	—	—	0.04
Mg	—	—	188 000
Al	—	—	16 000
Ti	—	—	800
Si	— (+)	—	225 000
Na	—	—	8 400
Cr	300	1 200	3 900
Fe	907 000	635 000	98800
Ni	88 000	1 000	100
Pb	0.2	12	1
Cu	200	500	1.5
As	12	10	0.3
Pt	8.4	0.7	1.6
Ru	5.2	0.03	0.3
Os	1.4	—	0.2
Ir	2.8	—	0.5
Au	1.5	0.1	0.005

(−) This signifies that the element is negligible in abundance in that phase as compared with the others.

(+) Some Si is found in the metallic phase in meteorites which have been subject to a high degree of reduction (the enstatite chondrites – see Section 3.6.2).

3 The affinities of a given element within this simple classification depend very much upon the degree of oxidation of the system. In addition to examples cited on p. 56 of *Principles of Geochemistry*, we can note that manganese is normally lithophile but in the most reduced group of meteorites – the enstatite chondrites (Section 3.6.2) – it occurs as a sulphide phase. Even magnesium and calcium, which are normally extremely lithophile, may be chalcophile under certain circumstances, for MgS and CaS have been discovered in some of the most highly reduced meteorites. Evidently we need to know details of the oxidation/reduction equilibria in a system before we can specify the behaviour of elements within the phases present.

4 In this classification, the siderophile elements are characterized by a weak affinity for oxygen and sulphur. They are readily soluble in molten iron and might therefore be expected to be preferentially enriched in a nickel-iron core. The chalcophile elements show a strong affinity for sulphur and occur as sulphides. These occur in mineral veins at the Earth's surface and in many crustal rocks as minor minerals, and it is possible that one or more sulphide layers occur within the mantle or core (Unit 2). The lithophile elements are enriched in the silicate layers of the Earth – the mantle and crust.

5 Note that Table 3.10 is based on data from meteorites and some elements may show different tendencies in the Earth. For example, Ga is concentrated in iron meteorites but occurs in silicates and sulphides in the Earth's crust.

3.3 The Chemical Basis of the Geochemical Classification

(*Objectives 1, 3*)

Why is it that we can set up a geochemical classification at all? What factors control the affinities of an element for a particular phase?

Table 3.11 in *Principles of Geochemistry* shows that elements of contrasted affinity occupy distinct regions of the Periodic Table. For example, the siderophile elements are largely among the 'transition' elements (S100[4]). There is obviously a relationship between the geochemical classification and the atomic properties of the elements.

This relationship is introduced on p. 57 of *Principles of Geochemistry*, where the relation between three atomic properties and the geochemical classification is described:

1 *Atomic volume*. A plot of atomic volume against atomic number and a discussion of the factors influencing atomic number and atomic volume are give in S100[5]. If you like, have a look at this and use it to amplify the comment made on p. 57 in *Principles of Geochemistry*.

2 *The heat of formation* of an element oxide. This is one measure of the ease of oxidation of an element. The importance of this property and its relationship to the geochemical classification and evolution of the Earth is discussed in Section 3.8.2.

3 *The electrode potential*. This is a measure of the tendency of an element to pass into solution in the form of its ions (i.e., the reaction $X \rightarrow X^+$). The more positive the potential the greater is this tendency. A list of elements arranged in order of decreasing electrode potential is known as the *electrochemical series*.

Since the geochemical classification can be related to these atomic properties, it must also be related to the fundamental structures of the elements.

Let us first consider the siderophile tendency. Siderophile elements tend to occur in the elemental state, which suggests that the valence electrons are not readily available for bonding purposes. A measure of the tightness with which an electron is held in an atom (i.e. its bond-forming ability) is the ionization energy or *ionization potential* (S100[6]).

Can you remember the definition of this?

The ionization potential (I) is defined as the energy required to remove an electron from an atom in the gas state. Since the element distributions with which we are dealing did not originate in the gaseous state, you should realize that the use of I is an oversimplification, although we can use it to make useful general predictions. If the ionization potential is high, much energy is needed to remove an electron and vice versa. Consider copper and sodium.

What are the important features in the atomic structures of these elements?

The atomic structures are represented below:

*Electron energy shells**

	1	2	3	4
Na	$1s^2$	$2s^2\ 2p^6$	$3s^1$	
Cu	$1s^2$	$2s^2\ 2p^6$	$3s^2\ 3p^6\ 3d^{10}$	$4s^1$

Both elements have a single valence electron outside a closed shell of electrons (8 in sodium, 18 in copper). The first ionization potential of sodium is 5.1 ev as compared with 7.7 ev for copper.

What do these figures tell us about each element?

The higher value for copper shows that its single valence electron ($4s^1$) is more tightly held than that of sodium ($3s^1$). This is presumably due to the less effective

* These are numbered using the principal quantum numbers. If you are unsure about the data in this Table look back to S100[7].

screening of the valence electron of copper from the nuclear charge, by the 18 electron shell.

Look back at Table 1. From the behaviour in the phases of meteorites, how could copper and sodium be geochemically classified? Does this relate to atomic structure?

Sodium is lithophile. This is consistent with its readiness to lose an electron or oxidize (S100[8], and this Unit, Section 3.8.2) and form ionic compounds. Copper, on the other hand, is dominantly chalcophile, although occurring naturally in small amounts as native metal and in very small amounts in silicate phases. This is explained by its high ionization potential and consequent preference for uncombined (siderophile) and covalent (sulphide) phases. Similar predictions can be made for the groups in which our examples occur – Group 1A for Na and 1B for Cu (S100[9]). The ionization potentials for these two subgroups are tabulated below. The features of the Group 1A elements have been discussed in S100[10].

Table 2 Ionization potentials

Group 1A	*Ionization potential* I_1(ev)	Group 1B	*Ionization potential* I_1(ev)
Li	5.3		
Na	5.1		
K	4.3	Cu	7.7
Rb	4.1	Ag	7.5
Cs	3.9	Au	9.2

In passing from Li to Cs (Group 1A), the ionization energy decreases. This is because the greater number of electron shells in Cs provides a more effective screening of the nuclear charge than in Li. The position is, however, reversed in Group 1B. Ionization potential is related to nuclear charge, ionic radius and the degree of screening afforded by the electron shells and the interrelations of these are not always simple (S100[9]).

From the data above, however, it might be predicted that the Group 1A metals are lithophilic and those of Group 1B chalcophilic or siderophilic. These predictions are basically sound (see Table 1). The alkali metals occur overwhelmingly in the silicate phase of meteorites and never occur as native metals or sulphides at the Earth's surface. Copper is found in the Earth's crust as native copper (uncommon), copper sulphides (predominant) and, in very small amounts, substituting for other elements in silicates. Gold occurs mainly in the native state, often alloyed with 5–15 per cent silver. Lesser amounts of gold occur as mixed sulphides, but gold is not found replacing metal ions in silicates. Silver behaves in much the same way as copper.

SAQ 2 The first ionization potentials of three unknown elements and iron are given below:

Element	*Ionization potential* I_1(ev)
A	7.63
B	3.89
C	8.96
Fe	7.90

(a) Put A, B and C in order of ease of oxidation relative to Fe, by matching these elements with the following statements:

(i) Much more easily oxidized.
(ii) Similar ease of oxidation.
(iii) More difficult to oxidize.

(b) Use this information to match each element with the description of its natural occurrence below:

1 Exclusively found in silicates.
2 Usually found native, i.e. as the uncombined element.
3 Rarely native, common as the sulphide and hydrated silicate.

We can summarize the basis for the geochemical classification of the elements as follows:

1 The lithophile elements are more easily oxidized than Fe^{2+}, because they have low ionization potentials and therefore readily lose electrons to form ions with the noble gas configuration, having 8 electrons in the outermost shell. The most important lithophile elements occur in groups I and II. These are the alkaline and alkaline earth elements. Other lithophile elements enter silicate compounds by forming bonds of largely ionic character.

2 The chalcophile elements are less easily oxidized than Fe^{2+} because they have higher ionization potentials than the lithophile group. Such elements have ions with 18 electrons in the most complete outer shell. These include elements in B sub-groups of older Periodic Tables – IB (Cu, Ag); IIB (Zn, Cd, Hg); IIIB (In, Tl); IVB (Pb); VB (As, Sb, Bi); and VIB (Se, Te) (see Table 3.11 in *Principles of Geochemistry*). These elements form strong covalent bonds and therefore stable compounds with sulphur. Note exceptions to this generalization. Some B sub-group elements are lithophile: for example, Ga (Group IIIB) behaves similarly to Al (Group IIIA) and substitutes for this major element in its minerals in the Earth's crust. Furthermore, some of the B sub-group elements can be found native, i.e. exhibiting siderophile tendency, the best examples being perhaps Cu and Ag.

3 The siderophile elements are heavier B sub-group elements which scarcely ever form covalent bonds with sulphur, let alone become oxidized. They occur at the Earth's surface as native metals and would be alloyed with the postulated metallic phase in the Earth's core. They include Ru, Rh, Pd, Os, Ir, Pt and Au.

4 A small number of elements remain, principally iron and closely related transition elements. They occur chemically combined in a number of ways, and may be found in oxide, silicate, sulphide and even native form. However, under the inferred conditions of the Earth's core they would all be reduced and alloyed with the postulated metallic iron phase, in which the bonding would be of metallic type. Such elements include Fe (of course), Mn, Co, Ni and S.

This is an oversimplification – but suffices for our purposes.

SAQ 3 Match the possible forms of occurrence of elements in list 1 with the four groups of the geochemical classification in list 2:

List 1	*List 2*
1 Form silicate minerals.	A Lithophile
2 Occur as sulphides in ore deposits.	B Chalcophile
3 Found as native metals at the Earth's surface.	C Atmophile
4 Occur as oxides in ore deposits.	D Siderophile
5 Form the atmospheric gases.	
6 May be alloyed with metallic iron in the core.	
7 Concentrate in the sulphide phase of meteorites.	

3.4 The Bulk Chemical Composition of the Earth

(*Objectives 1, 6*)

The model of the Earth that we have built up at this stage (Unit 2) has an iron-nickel core, a peridotite mantle and a crust with a chemical composition between basalt and granite. We can estimate the masses of the different parts of the Earth. This calculation is performed in *Principles of Geochemistry*, Table 3.1 (cf. S100[1]). We can also determine a geochemical classification of elements from meteoritic analogies (Section 3.2) and by applying these assumptions, the bulk chemical composition of the Earth can be simply determined.

chemical composition of the whole Earth

Read pp. 51–4 in *Principles of Geochemistry* (The composition of the Earth as a whole). In Table 3.8, the third column heading should be 'Table 3.7'. Replace the key to Table 3.9 with the following:

* From Tables 3.3 and 3.4.	‡ From Table 2.4.
† From Tables 3.7 and 3.8.	§ From Table 2.3.

Glossary for this reading

These terms are briefly introduced here and will be explained more fully in Section 3.6.

ACHONDRITE — A type of meteorite composed largely of silicate minerals and resembling terrestrial basalt.

CHONDRITE — The commonest type of meteorite, containing sulphide (troilite), metallic (iron-nickel) and silicate phases. The average chondrite represents the average of chondrite samples from the dominant groups.

PALLASITE — One type of a rare group of meteorites which contain approximately equal amounts of metallic (iron-nickel) and silicate phases.

STONY METEORITE — A meteorite consisting largely of silicate minerals. Such meteorites include the chondrites and achondrites.

Note the following points in this section of the set book:

1 Mason uses meteorite compositions as a basis for his calculations. He is here using the chondritic Earth model which we shall be discussing presently. Accept his assumptions for the moment, note the qualifying remarks he makes about the composition of mantle and core, and recall our discussion of these in the second half of Unit 2.

2 Mason favours the presence of sulphur rather than silicon as the density-reducing phase in the core. There is, of course, no direct way of deciding which alternative is more likely to be right!

3 In Table 3.8 the first four elements are the most abundant in Mason's, Washington's and Niggli's calculations, and the agreement is quite close. Thereafter, there are considerable differences but, as Mason says, not too much weight should be placed on the actual figures, considering the assumptions and uncertainties inherent in this kind of computation.

4 What is important is his point that 99.9 per cent of the Earth contains 15 elements, and 0.1 per cent contains the rest of the elements. This 0.1 per cent includes many of the elements extensively used in industry, such as lead, zinc, copper, uranium, gold, mercury and tungsten – we examine this aspect in more detail in Unit 5.

5 Finally it is worth noting again what a minute proportion of the whole Earth makes up the crust, the only part directly accessible to us (recall Unit 1, Section 1.3).

In Table 3.8 we are looking at estimates of the chemical composition of the Earth as it is today. But could it always have had this composition? It is very difficult to determine the nature and scale of past compositional changes. One process which must have caused a significant change in the Earth's chemical composition at an early stage in the Earth's history was described in S100[11].

Can you remember what it was?

The loss of the Earth's primitive atmosphere, which may have had an enormous mass and contained elements not present in today's atmosphere; this would certainly have appreciably changed the Earth's chemical composition.

To study the nature of possible changes in the Earth's chemical composition, we have to look at possible materials from which the Earth may ultimately have accreted. These are considered in the next Sections.

3.5 Solar and Stellar Spectra *(Objectives 1, 4, 6)*

In S100[12,13] you studied the broad features of the Solar System and the way in which these features may have originated. According to most theories on the origin of planets, the planetary material was derived from the Sun, or from a fairly homogeneous cloud of gas and dust which condensed to form both the Sun and planets. Study of the chemical composition of the Sun should thus provide valuable information on the chemical composition of the primitive Earth.

By way of a brief introduction to what follows, read pp. 15–16 in *Principles of Geochemistry* (The composition of the Sun). Note the limitations on obtaining this data and note also that the abundance figures are normalized to (i.e. expressed relative to) a conveniently chosen figure for Si.

chemical composition of the Sun

The composition of the Sun is essentially that of the Solar System since 99.86 per cent of the mass of the Solar System is accounted for by the Sun (all the planets account for about 0.14 per cent!) Solar abundances are thus useful in looking outward toward the composition of the Solar System, and in looking inward toward the bulk composition of the Earth. Many spectroscopic studies have been carried out on solar and stellar radiation. At least half the observing time of many large telescopes is spent studying spectra. You have read about the features of solar radiation in S100[14] and you have probably studied the solar spectrum with the OU spectroscope.

Can you remember the main features of the solar spectrum?

It consists of a continuous spectrum with thousands of superposed spectral absorption lines – the Fraunhofer lines. These are due to absorption of specific wavelengths within the solar atmosphere. Although the spectrum was first observed by Wollaston in 1802, it was Fraunhofer who first mapped out several of these lines in 1817. He labelled nine of the most prominent with letters of the alphabet (Fig. 1). Later, in 1882, H. A. Rowland photographed and published a 40 ft long map of the Sun's spectrum. Two small parts of this are reproduced in Figure 2. The dark lines matched to sodium and calcium are labelled with the same letters as in Figure 1, so you can see that this is a very detailed record indeed of the Sun's spectrum.

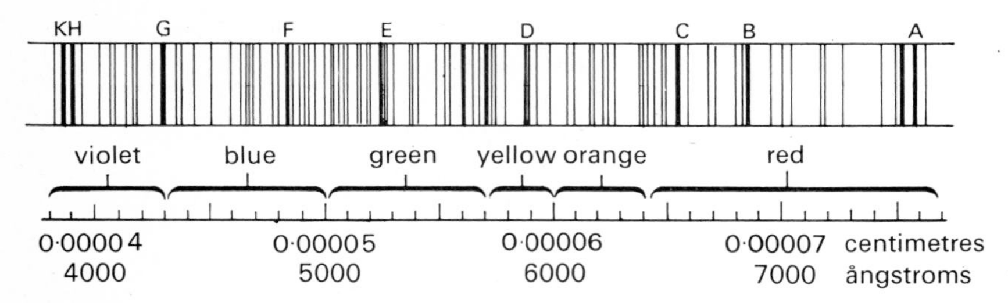

Figure 1 Diagram of the solar spectrum indicating the most prominent lines marked with letters of the alphabet as first labelled by Fraunhofer.
A, B=oxygen; C, F=hydrogen; D=sodium; E, G=iron; H, K=calcium.

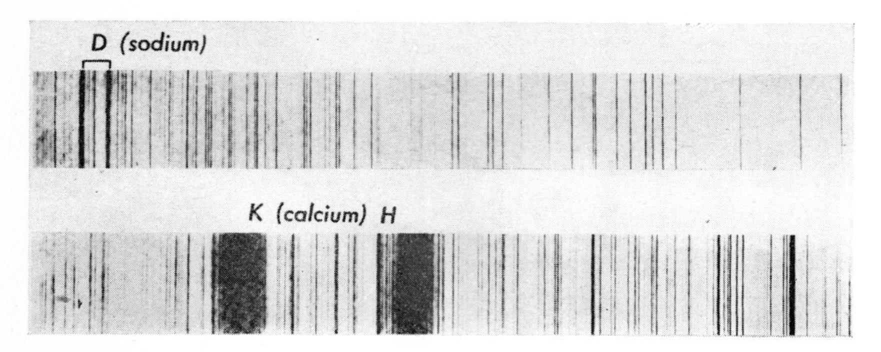

Figure 2 Two small parts of the solar spectrum from Rowland's original large-scale map. D=sodium; H, K=calcium.

The principles used in studying this spectrum are the same as those of absorption and emission spectrographic analysis (Unit 1, Section 1.7.2). However, in a spectrographic analysis, the unknown radiation is compared with that of a standard and great care is taken to ensure that the excitation conditions are not only appropriate for the particular analysis under consideration, but also reproducible for successive analyses. For solar radiation, by contrast, comparison with a standard is not possible, the appearance of spectral lines in the solar spectrum depending partly upon the concentration of the excited element and partly upon the physical environment of the solar photosphere. The analysis of solar spectra is rather like carrying out a spectrographic analysis on a heterogeneous sample, using uncontrolled excitation conditions and no internal standards at all! The consequently rather complicated approach to the determination of solar chemical abundances by spectroscopic techniques will now be outlined.

The measurement of the spectrum itself is a relatively simple operation. In practice, the absorption spectrum of the Sun and the emission spectrum of a laboratory source (usually iron) are recorded on to the same photographic plate. An example is shown in Figure 3.

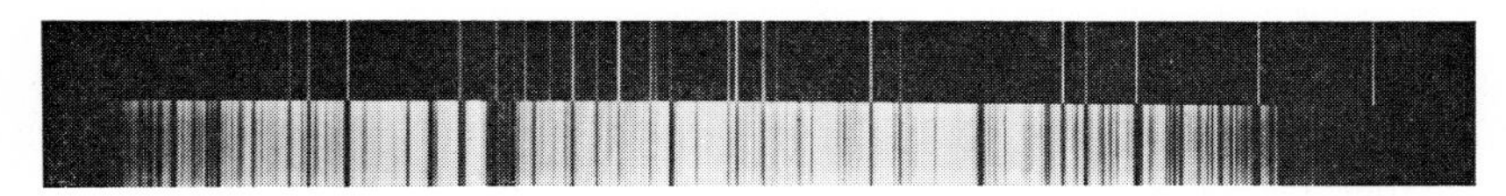

Figure 3 A portion of the solar (absorption) spectrum with a comparison (emission) spectrum of iron photographed with the same spectrograph.

Examine Figure 3 briefly.

Do the emission lines (light) coincide with any absorption lines (dark)?

Yes, they do, so which element do we immediately know must be present in the Sun? Iron, of course, since this is what was used for the laboratory source.

One could use all the elements as laboratory sources in the same way, but it is more economical to use the accurately known wavelengths of the iron emission lines to determine the wavelengths of the absorption lines (i.e. the wavelength of the radiation which has been absorbed to give the dark lines). Since each element has a characteristic radiation wavelength, the elements in the Sun can be identified.

By using spectrum photography, it has been possible to identify about two-thirds of the known chemical elements in the Sun. More elements are not found either because they are too rare to produce absorption spectra or, if present in sufficient quantity to give spectra, the temperature may be too high or too low to bring out the lines.

So much for qualitative data from solar and stellar spectra. To obtain quantitative analytical data, more manipulation is necessary. The spectra we observe all originate within the outer region of the Sun, the thin shell from which light escapes, called the *photosphere*. The first stage in calculating solar abundances involves using the overall features of the spectrum to estimate the temperature of the photosphere. These data, together with other information on the size and density of the Sun, are used to estimate a *model photosphere*, which represents a theoretically determined profile of pressure, temperature and density throughout the photosphere.

photosphere

model photosphere

Upon what two factors will the strength of an absorption line depend?

The strength of a single absorption line, measured in a spectrum, will depend upon the abundance of the relevant atomic species *and* upon the fraction of its atoms that are in the right state of ionization and excitation to produce the line. We have to separate these two factors if we are to relate the absorption line strength *only* to atomic abundance. Further complex calculations enable these effects to be separated. It is possible theoretically to predict how line strength varies with atomic concentration of any given element. The resulting relationship can then be used to determine the abundance of that element. This basic procedure has been used to determine element abundances not only in the Sun, but in many other stars as well.

Besides errors associated with the measurement of spectral intensities and the interpretation of the complex spectra, there is a further possible source of error: the surface of the Sun may not be representative of the Sun as a whole. Although this may appear severely to limit our knowledge of the Sun obtained in this way, it is, in fact, advantageous because the primitive (original) composition of the Sun cannot fail to have been modified by nuclear reactions taking place inside it (S100[15]). Because the near surface layers and the photosphere are cooler and have not been subjected to these processes, information gained from them is relevant to our search for the Sun's primitive composition, before it was modified by nuclear reactions.

Solar abundance estimates are very valuable. They provide the best source of data for the overall abundances of volatile elements within our Solar System. They also provide data on the cosmic abundances of nonvolatile elements which can be compared with other sources of cosmic element data, especially meteorites (see Section 3.6).

cosmic element abundances

Studies of this type have shown that over 95 per cent of stars have a composition essentially similar to that of the Sun. From 50 to 80 per cent of the mass of most stars seems to be hydrogen; H and He together make up from 96 to 99 per cent of the mass of all stars and, in some, more than 99.9 per cent.

SAQ 4 Which of the following components are contained in the solar spectrum?

(a) Continuous spectrum.

(b) Line emission spectrum.

(c) Line absorption spectrum.

SAQ 5 Very strong absorption lines with wavelengths about 6550Å and 4820Å are seen in a stellar spectrum. Look at Figure 1. This suggests that the star contains:

oxygen	hydrogen	iron	calcium	none of these elements
☐	☐	☐	☐	☐

SAQ 6 To determine quantitative element abundances from a stellar spectrum, it is necessary to (tick the boxes):

1 Determine a model photosphere. ☐

2 Refer the spectral lines to that of a star, the chemical composition of which is independently known. ☐

3 Estimate the proportion of atoms throughout the photosphere that are under the right physical conditions to produce an absorption line. ☐

4 Predict how line strength is related to atomic abundance. ☐

3.6 Meteorites *(Objectives 1, 5, 6)*

You have studied the main features of meteorites in S100[16]. The physical features of meteorites are discussed in TV programme 3.

Read pp 17–21 in Principles in Geochemistry. Don't worry about the details of mineral names and formula, or the numerous varieties of meteorite (except those listed in Table A), which are considered in more detail below.

classification of meteorites
Widmanstätten structure
meteorite finds and falls

Glossary for this reading

ASTEROID — A piece of planetary material from the belt of small fragments (ranging up to ~700 km in diameter) which circles the Sun between the orbits of Mars and Jupiter, and is known as the asteroid belt.

COSMIC RAYS — Extremely fast moving particles (believed to be electrically charged atomic nucleii) which are continually entering the Earth's upper atmosphere from interstellar space.

EXSOLUTION — Unmixing of a single homogeneous solid into two homogeneous solids in response to a change in physical conditions.

OBSIDIAN — A glassy volcanic rock formed by such rapid chilling of acid magmas that crystals do not form.

PLANETESIMAL — One of the small bodies of planetary material which was assumed to have accreted to form the primitive Earth. Such bodies may have been similar to the asteroids.

RHYOLITE — A volcanic rock of acid composition, equivalent in chemical composition to granite. (Obsidian would be a very rapidly cooled rhyolite.)

Now do the following questions which cover the important parts of this passage:

SAQ 7 Summarize the characteristic features of the four principal meteorite types by matching the names with appropriate letters.

1 Chondrites
2 Achondrites
3 Stony irons
4 Irons

A consist essentially of an Fe-Ni alloy.
B contain Mg-Fe silicates.
C contain approximately equal amounts of Fe-Ni alloy and Mg-Fe silicates.
D contain chondrules.
E contain little or no Fe-Ni alloy.
F resemble terrestrial igneous rocks.
G contain 10–20 per cent Fe-Ni alloy.

SAQ 8 (a) Can you explain why the group of meteorites most frequently seen to fall (the falls) is not the same as the group most frequently recovered at the Earth's surface (the finds)?

(b) What is the principal difference between the eight most abundant elements in chondrites (Table 2.4 in *Principles of Geochemistry*) and those in the Earth (Table 3.7).

3.6.1 Meteorites and the composition of a primitive Earth

Although the Sun provides the best information about the abundance of both the volatile and non-volatile elements in the Solar System because of its enormous bulk, you will recall that the Sun consists largely of H and He, and that only about two-thirds of all known elements have been recorded in its spectra.

In order to seek a possible composition for our primitive Earth, we need a better estimate for the non-volatile composition of the Solar System – clearly meteorites offer us the best chance of getting this, but which ones?

You now have information about four main kinds of meteorite:

1 Irons
2 Stony irons
3 Chondrites
4 Achondrites

We need to select one of these as representative of the non-volatile material in the

Solar System. You will notice that we are making two prime assumptions here:

(a) The Earth was originally homogeneous and became segregated into layers (Section 3.1, alternative 2).

(b) Meteoritic material represents planetary material, *similar* to the Earth, which has been fragmented. We assume that meteorites provide us with samples of bodies which have differentiated from a primitive homogeneous stage into layered structures like the Earth. We also assume that some meteorites represent the primitive homogeneous stage and are made of material similar to that which accreted to form our own Earth. Our task is now to find which of these meteorite groups fulfils the latter requirement best. Let us look again at the four groups of meteorites above. Recall their definitions and consider their suitability as primitive Earth material, bearing in mind the information we already have about the Earth, comparing the *physical* nature of the core and the physical *and* chemical nature of the mantle and crust.

Obviously the *irons* could not represent the primitive material from which our Earth was formed, for they contain virtually no silicate – they presumably represent the core region of some other planetary body. The *stony iron* group are not much more suitable, since they too contain insufficient silicate – they could be inferred to represent perhaps a deep mantle/core transition region. The *chondrites*, on the other hand, are eminently suitable, as they have a moderate amount of nickel-iron, significant sulphide, and abundant silicate, including feldspar as well as olivine and pyroxene. The *achondrites*, finally, resemble terrestrial igneous rocks, those of basaltic composition in fact, and are without a metallic phase. This makes them totally unsuitable as primitive Earth materials: one could envisage them as representing the former crustal regions of a now disrupted planetary body. Clearly we need to concentrate our attention on the chondrites. Before we do so, answer *SAQ* 9.

SAQ 9 Meteoric sulphur occurs chiefly in troilite (meteorite iron sulphide – FeS). In iron meteorites this is found as inclusions (e.g. the Rowton iron meteorite seen in TV programme 3), which may be several centimetres in diameter. In chondrites, the troilite often occurs as very much smaller particles scattered throughout the matrix. Think back to Unit 1 – would it be easy to take a representative sample for the analysis of sulphur from (a) an iron meteorite, (b) a chondrite? What other group of elements might occur in the troilite phase and hence be subject to the same sampling considerations?

3.6.2 Chondrites

The chondrites comprise over 80 per cent of the meteorites seen to fall. They form a closely related group, chemical variations within which are in general rather small. Chondrites are characterized (with a few notable exceptions) by the presence of *chondrules*, small rounded structures composed principally of olivine, sometimes with pyroxene as well.

chondrites
chondrules

SAQ 10 Have a look at Figure 4. This shows thin sections of four meteorites: two chondrites and two achondrites. Study the picture and indicate the meteorite types by marking the boxes below.

	A	B	C	D
chondrite				
achondrite				

Paradoxically, not all the meteorites classified as chondrites contain chondrules. Some chondrites have been heated since their origin. Since heating has the effect of recrystallizing the minerals, this may lead to the obliteration of chondrules (Fig. 5).

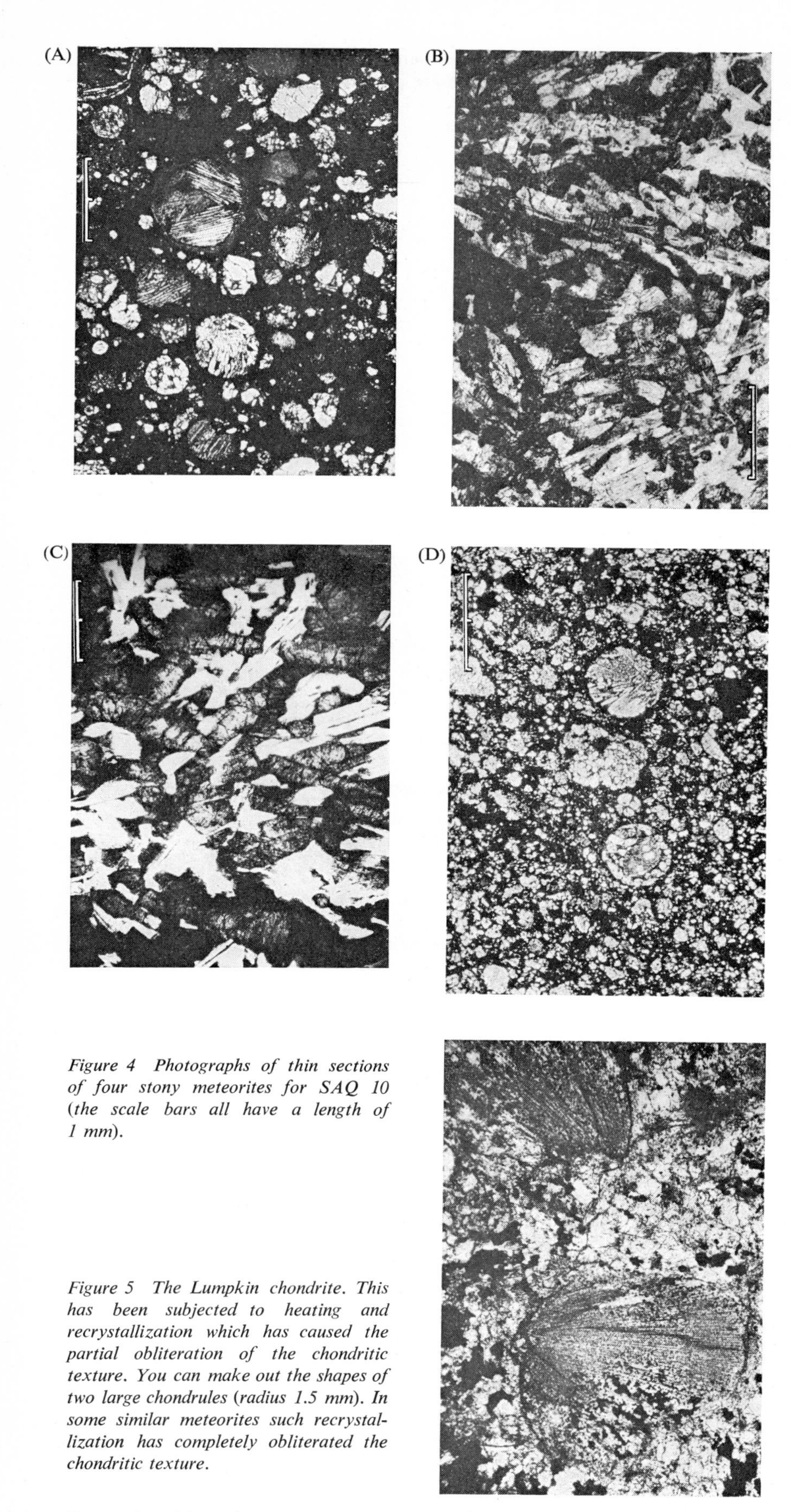

Figure 4 Photographs of thin sections of four stony meteorites for SAQ 10 (the scale bars all have a length of 1 mm).

Figure 5 The Lumpkin chondrite. This has been subjected to heating and recrystallization which has caused the partial obliteration of the chondritic texture. You can make out the shapes of two large chondrules (radius 1.5 mm). In some similar meteorites such recrystallization has completely obliterated the chondritic texture.

Some chondrites, however, seem never to have contained chondrules; but, because the chemical and mineralogical features of chondrites are rather uniform, in the very small number of cases where chondrules are not developed it is not difficult to classify the meteorite correctly. Chondrule-free chondrites are very important and some of these will be discussed presently.

Five major groups of chondrite, distinguished on the basis of mineralogical and chemical features, are classified in Table 3, together with the total number of specimens (falls and finds recorded up to 1966).

chondrite groups

Table 3 The major groups of chondrites

Group	*Variety*	*Number of falls and finds*
1 Enstatite* chondrites	Type I Type II	7 8
2 High-iron chondrites (H)		~400
3 Low-iron chondrites (L)		~500
4 Low-iron, low-metal chondrites (LL)		~50
5 Carbonaceous chondrites	C3 C2 C1	10 16 5

* A variety of pyroxene ($Mg_2Si_2O_6$)

These groups are distinguished by a variety of chemical and mineralogical features. For example, the enstatite chondrites are characterized by the occurrence of the magnesium pyroxene, enstatite, and the absence of olivine. The two most abundant groups (H and L) are distinguished by their differing total iron contents. Despite their large abundance, these *ordinary chondrites* are not believed to be representative of the primitive non-volatile element abundance we are seeking. For this purpose, the C1 group of *carbonaceous chondrites* has been selected; the reasons are outlined in the following Sections.

ordinary chondrites

C1 carbonaceous chondrites

3.6.3 Cosmic abundance patterns

The processes which created the different nuclides (S100[17]) are not known in detail. However, a relationship may exist between the mass numbers, A, (S100[18]) and the cosmic abundances of nuclides. Read pp. 21–5 of *Principles of Geochemistry* (The cosmic abundance of the elements).

cosmic abundance patterns

Oddo-Harkins rule

Read this passage now and answer the following questions by reference to Figure 2.1 in the book:

SAQ 11 What is the difference between the cosmic abundance of C and Ga? Explain it.

SAQ 12 Mark the following statements true or false:

	T	F
(a) Elements of low atomic numbers are generally more abundant than those of high atomic number.		
(b) Elements of even atomic number are generally less abundant than those of odd atomic number.		

The important point about cosmic abundances is that there are considerable regularities in the graph of element abundance against atomic number. Although the graph in Figure 2.1 has been drawn using non-volatile element abundances obtained from ordinary chondrites, of all the meteorite types, the C1 chondrites show the *clearest* relationship between A and concentration (Fig. 6).

Figure 6 shows that the relationship is complex, for the concentration curve rises to several peaks. Taken as a whole, however, the curve is smoother than could be produced by using any other of the chondrite groups. Because the C1 chondrites show the smoothest relationship between A and abundance we think that their composition has special significance – we think that the C1

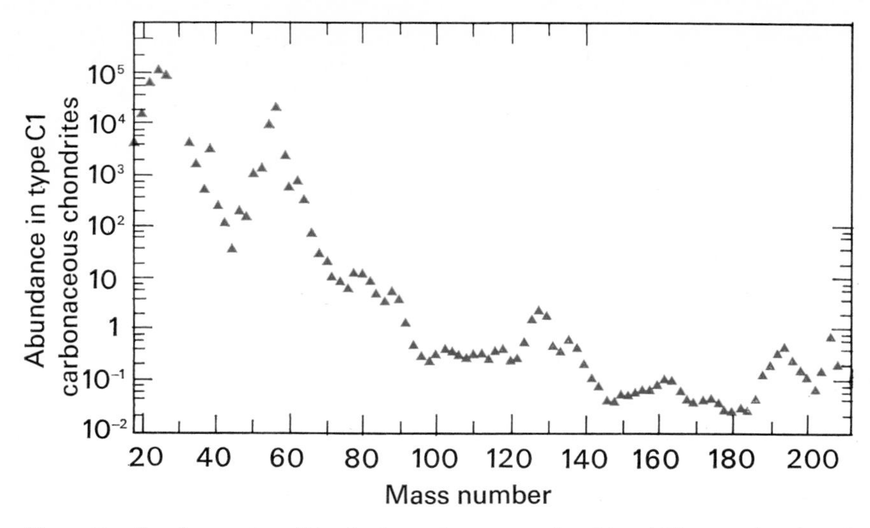

Figure 6 Abundance of nuclides in the carbonaceous chondrites (C1) as a function of mass number. The pattern is the same whether odd- or even-A nuclides are used; for simplicity, therefore, only odd-A nuclides have been plotted. Abundances are relative to Si=10^6 atoms in the meteorites (cf Fig. 2.1 in Principles of Geochemistry where abundances are relative to Si= 10^4 atoms; the difference does not affect the abundance patterns).

chondrites provide the closest approach to cosmic abundances among the meteorites.

3.6.4 Trace element abundances in chondrites

Many elements vary in abundance between the groups of chondrites listed in Table 3. Several trace elements, for example, are depleted in ordinary chondrites relative to the C1 group. Table 4 compares the abundances of elements in ordinary chondrites with those in C1 chondrites.

Table 4 Abundances of elements in ordinary chondrites compared with two samples of C1 carbonaceous chondrites

Element	*Relative abundance**
Fe, Si, Mg, K, Sc	Not significantly different from 1
Rare earths, Ti, Cr, Mn, Co, Ba	0.8–1.0 Depletion probably not significant
Na, Rb, Cs	0.4–0.7
Ca, Sr, Y, Al, U, Th	0.3–0.8
P, V	0.2–0.4
F, Cl	0.1–0.3
S, Te	0.1–0.2
Cu, Zn, Ge, Pb	0.1–0.3
Bi, Tl, Hg	0.001–0.1
C, H, N	0.005–0.01

* Relative abundance is expressed as:

$$\frac{\text{Atoms \% element in ordinary chondrites/Atoms \% Si}}{\text{Atoms \% element in Type 1 carbonaceous chondrites/Atoms \% Si}}$$

Element abundances in Table 4 are again normalized to Si (as in Fig. 6, and in Figure 2.1 of *Principles of Geochemistry*) so that the relative abundance patterns are not affected by variable proportions of nickel-iron and troilite in different meteorites.

SAQ 13 Using Table 4, place the following four elements in order of increasing depletion (relative to Si) in ordinary chondrites relative to C1 carbonaceous chondrites:

Cr
Cl
Hg
Rb

The enrichment of so many minor and trace elements in C1 chondrites relative to the other chondrites is evidence that they themselves have not undergone much change and that they are accordingly more primitive than the other groups. The next Section explains why.

3.6.5 The significance of chondrules

Although most chondrites (over 98 per cent) contain chondrules, these structures are missing in some types, especially those in which the chondrules have been obliterated by recrystallization caused by reheating (Fig. 5). In the C1 chondrites, however, there is no sign that chondrules have ever been present. None the less, their chemical similarity to the chondrule-bearing C2 and C3 carbonaceous chondrites makes it clear that the C1 meteorites belong with the chondrites.

significance of chondrules

So what is the significance of chondrules? Analyses of chondrules separated from the matrix in which they are situated (Fig. 4, A and D, shows how these two components can easily be distinguished) consistently show them to be depleted in about 20–30 elements, relative to the matrix.

This depletion is presumably related to the chondrule-forming process, which is poorly understood (not surprisingly, for nobody has actually seen a chondrite form!), but it may involve heating and partial melting in the vacuum of outer space, whereby small molten droplets are formed and then quenched. Heating of this kind in any rock will cause redistribution, or *fractionation*, of elements between solid and liquid phases.

trace element fractionation in chondrites

At all events, the presence in ordinary chondrites of chondrules, their depletion in some elements relative to the matrix, and the overall depletion of many elements relative to C1 carbonaceous chondrites (Table 4) combine to suggest that ordinary chondrites have been through a heating and fractionation process.

In contrast, C1 chondrites are not only unfractionated with respect to ordinary chondrites, they are also free of chondrules. (We should note that both C2 and C3 carbonaceous chondrites do carry chondrules.) In other words, C1 chondrites carry very little evidence of having been heated, and for these two reasons they are regarded as being compositionally the most primitive meteorite group.

How do they rate as the most primitive non-volatile matter of the Solar System?

3.6.6 Comparison of chondritic and solar abundance patterns

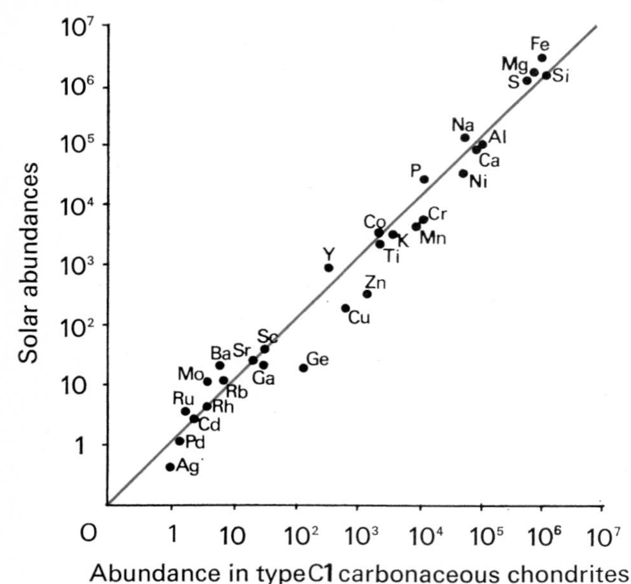

Figure 7 Comparison of element abundances in carbonaceous chondrites (C1) with those in the Sun, normalized to Si=10^6 atoms in the Sun and in the meteorites.

The overall chemical features of C1 chondrites match solar abundances very closely. Look at Figure 7. This shows a comparison of elemental abundances

chondritic and solar abundance patterns

(normalized to Si) between C1 chondrites and the Sun, as determined by spectroscopic studies. If the element abundances relative to silicon were identical in these two sources, then the abundances would plot on the straight line. The close approach to this straight line suggests that the C1 chondrites approximate the chemical composition of the primordial non-volatile matter of the Solar System. However, because of the low accuracy of solar data, some other minority chondrite groups (for example, C2), but *not* ordinary chondrites, also give a satisfactory match. This approach therefore, excludes ordinary chrondrites, but is not wholly conclusive in the selection of C1 chondrites as best representing the cosmic abundances of non-volatile elements. We need to look a little more closely at this group.

3.6.7 The C1 chondrites

features of C1 chondrites

Unfortunately we have only five samples of C1 chondrites. These are hardly sufficient to be representative of their parent material, but all have very similar features which are consistent with an origin at fairly low temperatures (below 500 °C). Under these conditions, reduction was evidently less pronounced than

Figure 8 A piece of Orgueil C1 carbonaceous chondrite. White streak in middle of specimen is a vein of magnesium sulphate.

in other meteorites, for there is no metallic phase. The iron is present as oxides and sulphides, while silicates occur as hydrated forms, as for example serpentine or chlorite.† A complex array of organic substances is present. These substances appear to have acted as 'glue' so the C1 chondrites have the appearance of coaly lumps (Fig. 8). The compositional features are summarized in Table 5.

Table 5 Mineral composition of the Orgueil (C1) carbonaceous chondrite

~63 per cent hydrated Mg-Fe layer silicate (probably chlorite)
~ 7 per cent magnesium sulphate (epsomite, $MgSO_4.7H_2O$)
~ 6 per cent magnetite (Fe_3O_4)
~ 5 per cent troilite (meteoritic iron sulphide, FeS)
~ 3 per cent gypsum ($CaSO_4.2H_2O$)
\+ minor elemental sulphur, S
sodium sulphate (Na_2SO_4)
organic materials (fatty acids, porphyrins, nucleic acids, amino acids, hydrocarbons)
\+ a variety of other substances of lesser importance

The survival of such a complex array of hydrous and organic materials is by itself quite sufficient to confirm that C1 chondrites cannot have been subjected to any significantly high temperatures – indeed, it is surprising that compounds so sensitive to dehydration as epsomite and gypsum survived the frictional heating which accompanied passage through the atmosphere. No wonder these meteorites are so rare.

What kinds of change would be involved in developing ordinary chondrites from C1 carbonaceous varieties?

Clearly, the first necessity is heating, to drive off the water and turn hydrous minerals into anhydrous ones. The silicates would thus be converted from hydrated mica-like minerals into olivine and pyroxene. As water left the system it would also remove some of the more easily fractionated components, particularly many of the minor and trace elements, leading to depletion in these constituents. Moreover, the departure of water, coupled with the effect of heating upon the hydrocarbon compounds and sulphur, would cause reduction of the system. Sulphate would become sulphide, and oxide would be reduced to metal, especially the iron oxides.

SAQ 14 Indicate whether the following statements about C1 carbonaceous chondrites are true or false:

	True	*False*
1 Large proportion of observed meteorite falls.		
2 Do not contain chondrules.		
3 Contain metallic iron.		
4 Show a clearer relation of mass number A to abundance than ordinary chondrites.		
5 Consist largely of Mg-Fe silicates.		
6 Have similar relative element abundances to those of non-volatile elements in the Sun.		
7 Are not depleted in elements as are ordinary chondrites.		

3.6.8 The chondritic Earth model

In the preceding Sections we have suggested that the C1 chondrites represent the material from which the other chondrites were ultimately derived. By analogy, the Earth may also be derived from material with the ultimate composition of C1 chondrites. Clearly, however, modification of C1 chondrite material was common – there are very many more 'ordinary' (H and L types, Table 3) than C1 chondrites. Since these modification processes seem to have operated widely among the parent meteorite planets, and since the ordinary chondrites are similar in composition to the whole Earth (i.e. mainly Mg-Fe silicate +Fe-Ni metal), many comparisons have been made between them and the Earth. The hypothesis that the Earth has a bulk composition close to that of the ordinary chondrites is known as the *chondritic Earth model.*

chondritic Earth model

Although the variety among the chondrites is limited, it is sufficient to suggest that they are unlikely to have originated as parts of a single parent body. Moreover, although the Earth may broadly resemble the majority of chondritic meteorites, these do not form a sufficiently homogeneous group for detailed comparisons to be made. The chondritic Earth model therefore provides a hypothesis for discussion and testing, rather than a basis for precise comparison – as you can see from the answer to *SAQ* 8 (b): estimated average chondrite and whole Earth compositions show significant differences.

So far, then, we have arrived at some plausible starting materials for a primitive Earth, namely the chondritic meteorites. Before passing on to consider how the Earth might have evolved from such materials, we should look at other extraterrestrial material which has recently become available to us for the first time: the lunar rocks. Do these throw any additional light on our problem?

3.7 Lunar Rocks (*Objectives 1, 7*)

Before 1969, meteorites were the only samples of extraterrestrial material available for study on Earth. However, since the Apollo Moonshots began, many samples of the lunar surface have been studied in great detail in laboratories all over the world.

The results of these studies could be important in our search for the composition of the primitive or primordial matter of the Solar System. For example, some rocks from the Moon's surface may be representative of the Moon as a whole. Alternatively it may be possible to deduce the internal composition of the Moon from surface samples, and it has also been suggested that some meteorites may have originated on the Moon. As you read this Section, which discusses these possibilities, bear in mind our earlier consideration of sampling problems. *Very* small areas of the Moon's surface have been sampled. How representative can such minute samples be, even of a comparatively small body such as the Moon? One possibility is that, if samples from widely spaced sites have similar features, they may be representative of a significant part of the Moon. It seems, fortunately, that this may be the case.

lunar regolith

At each base, the astronauts landed upon the Moon's 'soil' or *regolith* – a layer of rock fragments and dust – and collected rocks and finer materials. These are classified into 4 groups as follows:

Type

A Fine-grained igneous rocks analogous to basalt.

B Medium-grained igneous rocks analogous to dolerite (Unit 2, Appendix 2).

C Microbreccias – fragmented igneous rocks of types A and B broken by dynamic shock effects, believed due to meteorite impact.

D Fine-grained loose particles from the regolith.

The proportions of these constituents vary at the different landing sites but all types are always present. In general, basaltic rocks and their brecciated derivatives (Types A to C) have dominated at all the sites.

lunar anorthosite

The fine-grained material (Type D) contains a variety of particles. In a representative sample of 'soil', from one of the Mare landing sites studied by one group of investigators, 95 per cent of the particles were of basaltic type, 6 per cent were of a rock type called *anorthosite*, and 1 per cent were of other particles – which included broken fragments of meteorites which have pounded the Moon's surface. Anorthosite is a rock type, uncommon on Earth, which is composed essentially of crystals of plagioclase feldspar, and the rock is named after the calcium-rich plagioclase - anorthite. The lunar anorthosites are composed essentially of anorthite and differ from most terrestrial anorthosites in which the plagioclase composition is usually intermediate between albite (sodium plagioclase) and anorthite (see Unit 2, Section 2.7.7).

The abundant basaltic and doleritic fragments have probably come from the bedrock below the Mare landing site, while the anorthositic fragments have probably come from the lunar highlands – the most recent Apollo missions appear to have confirmed this.

The theory that some of the lunar highlands are of anorthosite composition is supported by three other lines of evidence: (i) the lighter colour of the highlands (anorthosite is very much lighter in colour than basalt); (ii) their lower gravity when compared with the Mare; (iii) the approximate chemical compositions of highland material analysed by soft-landed Surveyor spacecraft. These chemical analyses, although crude, are very similar to the more precise and accurate analyses that have been made of anorthosite fragments from the lunar soil.

In short, therefore, two major rock types have been found in the samples returned from the Moon. How similar are these types to their equivalents on Earth? In Table 6 are analyses of lunar basalts and anorthosite.

Table 6 Analyses of lunar and terrestrial rocks

	1	2	3	4	5
SiO_2	40.6	41.0	49.0	44.0	43.9
TiO_2	10.3	3.5	1.0	0.0	0.2
Al_2O_3	10.7	11.3	18.2	34.6	32.9
FeO**	18.8	20.2	8.9*	1.1	1.8*
MnO	0.39	0.24		0.0	0.0
MgO	8.0	12.0	7.6	1.3	2.7
CaO	9.9	10.2	11.2	18.3	16.6
Na_2O	0.55	0.46	2.6	0.7	1.4
K_2O	0.14	0.17	0.9	0.1	0.1
	99.4	99.7		100.1	99.6

1 Average of 8 rocks from Apollo 11 landing site.
2 Average of 22 rocks from Apollo 12 landing site.
3 Typical terrestrial basalt (Unit 1, Table 3).
4 Anorthosite fragment from Apollo 11 soil.
5 Anorthosite, formed by sinking of plagioclase crystals in a basic intrusion on the Isle of Rhum, Scotland.

* Iron all calculated as ferrous oxide, FeO.

** The lunar rocks have been formed under more strongly reducing conditions than those of the Earth. Moreover, there is no atmosphere on the Moon, so that oxidation cannot take place at the lunar surface. Hence, all the iron in the lunar rocks is reduced and is present as FeO rather than as Fe_2O_3 or Fe_3O_4. There is also very little H_2O in samples of Moon rock.

Compare the averages of the basaltic rocks from the two landing sites with the average terrestrial basalt.

What distinctive features do the lunar basalts have?

The lunar basalts have low contents of SiO_2, Al_2O_3, Na_2O, K_2O and high contents of FeO and TiO_2. The TiO_2 content of average 1 is spectacularly high in relation to any terrestrial analogue.

Although the basalts consist of similar minerals to terrestrial basalts (mainly pyroxene and plagioclase with olivine) they contain larger amounts of ilmenite†, accounting for the high TiO_2 contents in the analyses.

Now look at the chemical analysis of the lunar anorthosite. This is very similar to the chemical analysis given of a terrestrial anorthosite.

Which three oxides bulk largest in these chemical analyses?

Anorthosites consist essentially of SiO_2, Al_2O_3 and CaO as you would expect, for these are the constituents of plagioclase feldspar.

If basalt, dolerite and anorthosite represent the most important rock types on the lunar surface – what relationship do they bear to the Moon as a whole? Basalts and anorthosites are not really abundant in the Earth as a whole, being confined to the crust, and we might suspect that this would also be true of the Moon. In other words, the Moon may have a crust, just like the Earth, overlying an interior of different chemical composition.

In Unit 2 we discussed the effect of pressure on rocks of basaltic composition.

What happens to the minerals of basalt at high pressures?

At high pressures within the Earth, the mineralogy of basalt changes to produce a rock type of greater density than basalt—eclogite. Might the same change be expected for lunar rocks? Yes. Experiments carried out upon lunar samples have shown that the basaltic rocks transform to a higher density 'lunar eclogite' at pressures equivalent to a maximum depth of 300 km in the Moon (the radius of the Moon is 1 728 km). Besides containing garnet and pyroxene, as in terrestrial eclogites, the lunar eclogite would also contain ilmenite.

The density of the Moon as a whole has been determined as 3 340 kg m^{-3}. The surface basalts have a density of 3 300 kg m^{-3} (which is, incidentally, considerably higher than terrestrial basalt, whose density is around 3 000 kg m^{-3}).

SAQ 15 The lunar interior could be composed of material with: *Yes* *No*

1 A similar density and chemical composition to lunar basalt?

2 A similar density and different chemical composition to lunar basalt?

3 A different density and different chemical composition to lunar basalt?

Obviously the interior must be something with a similar density to lunar basalt, but if it were of similar composition it would be eclogite and too dense for the measured values.

The other rock type at the Moon's surface – anorthosite – can also be excluded as a significant component of the lunar interior because, under pressure, it also changes to a garnetiferous rock, denser than the Moon as a whole.

Investigators of the lunar rocks have suggested several alternatives for the lunar interior (or mantle). These include varieties of peridotite, which have the appropriate density – and we have the Earth as a precedent for inferring peridotite to underlie a basaltic crust. But there is no positive evidence yet.

Samples so far collected from the Moon's surface, therefore, are most unlikely to be chemically representative of the Moon as a whole, even though they may be reasonably representative of its surface. For the present, that is about as far as we can go.

The data provided by the Moon rocks has enabled the hypothesis of a lunar origin for meteorites to be examined closely. There are many chemical differences between chondrites and the lunar rocks. For example, Al is higher in *all* the lunar rocks analysed than in *any* chondrite (compare Table 6 and Table 2.4 in *Principles of Geochemistry*). It thus appears most unlikely that chondrites could be derived from the surface of the Moon.

Can you remember which meteorite group most nearly resembles basaltic igneous rocks?

Some achondrites are very similar in chemical, mineralogical and textural features to lunar basalts. Some members of this meteorite group could therefore have originated on the Moon's surface. It is equally possible, however, that achondrites have an origin as igneous rocks upon the surfaces of small asteroidal bodies. After all, both the Earth and Moon have basaltic rocks as major crustal components, so other planetary bodies could well have the same. An origin for achondrites, along with other meteorite types, in the asteroid belt is preferred to a lunar origin at the present time.

Summary

Although the lunar rocks represent a unique source of scientific data of tremendous interest, these samples tell us little about the composition of the primitive materials of the Moon, because we do not yet know what the lunar interior is made of. They do not help us, therefore, in our study of the chemical composition of the Solar System. The features of the Moon rocks are not consistent with their being a source for chondritic meteorites.

For the present, then, the Moon offers us no help in our quest for the primordial non-volatile matter of the Solar System—so we still have only the chondritic meteorites to go on. The next Sections are devoted to further discussion of the Earth's evolution from such material.

SAQ 16 Explain, in not more than two sentences, why samples retrieved from the surface of the Moon cannot be chemically representative of the Moon as a whole.

3.8 The Chemical Composition of the Primitive Earth

(*Objectives 1, 8, 9, 10*)

The best estimate of overall element abundances in the Solar System can be determined from the Sun's spectrum. The C1 chondrites are believed to provide the best estimate of the primordial non-volatile material of the Solar System. The ordinary chondrites and the Earth may have been derived from material with this original composition. (The same reasoning can presumably be applied to the Moon and other planets.)

Few of the Earth's physical and chemical features appear to resemble those of the C1 chondrites, however. There are more similarities between ordinary chondrites and the Earth, but again there is a major difference: metal, silicate and sulphide are mixed together in these meteorites, not neatly segregated as in the Earth. Clearly the chondritic parent bodies were not differentiated into a crust, mantle and core, and the denser Earth (5 500 as against 3 500–3 700 kg m^{-3} for chondrites) probably contains more iron relative to silicon than do the chondrites (cf. your answer to *SAQ* 8 (b)).

What processes have resulted in the present distribution of elements within the Earth?

3.8.1 The physico-chemical development of the modern Earth

Read pp. 59–65 in Principles of Geochemistry (The pregeological history of the Earth).

Glossary for this reading

CHERT — A compact siliceous rock of organic or chemical origin.

OCCLUDE — To absorb and retain a gas.

PARTIAL PRESSURE — For a mixture of gases contained in a fixed volume, the pressure that one gas would exert if it filled the whole of the volume is called its partial pressure.

SALT DOME — Salt is deposited in horizontal evaporite beds and, because it is less dense than common rock types, it tends to move plastically and rise into the overlying rocks as intrusive masses. These are called salt domes.

There are some important points to notice in this extract:

1 The evidence from heavy inert gases favours accretion of solid particles rather than condensation of incandescent gas.

2 Chondritic meteorite material, comprising metal, sulphide (troilite—FeS) and silicates is assumed to have formed the accreting particles.

3 Following accretion, the main heating effect came from radioactive sources. Radiogenic heating was much more vigorous in early stages than later.

radioactive heating of the Earth

SAQ 17 Why has the proportion of ^{40}K* and ^{235}U diminished so much through geological time, in comparison with ^{238}U and ^{232}Th?

SAQ 18 Study Figure 3.10 in *Principles of Geochemistry*, look at the text, and answer the following question:

At a time 600 million years after accretion, was the inferred temperature of the Earth higher than that of the melting point of (a) iron, (b) diopside (a pyroxene, $CaMgSi_2O_6$)?

4 It was this intense early radiogenic heating which initiated the process of core formation. The situation is rather like that in a blast furnace when molten iron and silicate slag have formed: the iron is denser and sinks to the bottom. So, in the Earth, denser iron sank to the centre—although admittedly the silicate 'slag' of the Earth was never completely molten.

formation of the core

The separation of the core by such a process involved the movement of vast masses of iron deep within the Earth. Such an immense internal reorganization would have far-reaching consequences, which probably included the nucleation of continents at the Earth's surface (Unit 5).

5 The discussion of crustal evolution considers two hypotheses. In one of these, an early *basaltic* crust was initially formed and was followed by the development of sialic crustal nucleii with the formation of island arcs (you will read more about this in Unit 5). The second model holds that the first crust to form was of granitic or granodioritic composition. In either case, it is likely that the early sialic crust was much thinner and less extensive than now, and has been growing by accretion from below and on the sides, ever since. It is important to note that some sialic rocks from Western Greenland have recently (1971) been radiometrically dated at 3980 Ma. This means that the age of the earliest sialic crust can be pushed back considerably from the figure of about 3.5×10^9 Ma given in *Principles of Geochemistry*.

6 pp. 63–5. You should realize that the origin and growth of the atmosphere and of life are stages in the chemical evolution of the Earth. You have studied these subjects in S100[11] and they will be referred to in later Units, but you should read these pages now to get a complete picture, without worrying too much about the details.

The early accretion

Can you spot the difference between the accreting material proposed in this section of *Principles of Geochemistry* and the nature of the meteoritic material we selected as being the most plausible primitive Earth material?

In the extract you have read, the accreting material was suggested to be similar to chondritic meteorites, containing metallic iron. The primitive material which we selected was similar to C1 carbonaceous chondrites which contain very little or no metallic iron.

There are two ways in which this apparent conflict can be resolved:

I Two-stage origin. Primitive oxidized material, similar to C1 chondrites, was heated and partially reduced *before* accretion. The Earth would then have formed by accretion of particles similar to those occurring in ordinary chondrites.

two-stage origin for the Earth

II Single-stage origin. It is also possible that the Earth accreted directly from oxidized (C1 chondrite) material, and that reduction of oxide to metal, loss of volatiles, melting and differentiation occurred through heating as a direct result of this primary accretion process.

single-stage origin for the Earth

Hypothesis I is the most popular and the one you have just read about in *Principles of Geochemistry*. Hypothesis II has been developed by Professor

* *In this Course we shall refer to isotopes with the same notation as in S100*[18]*, although this differs from that in* Principles in Geochemistry.

Ringwood of the Australian National University and is described below for comparison.

oxidation state of chondrites

Since these theories vary principally in their assumption of the initial oxidation state of the Earth, let us study first the degree of oxidation of the chondrites. *Look at Figure 9, which shows the relative proportions of oxidized (silicate) and reduced (metal and sulphide) iron.*

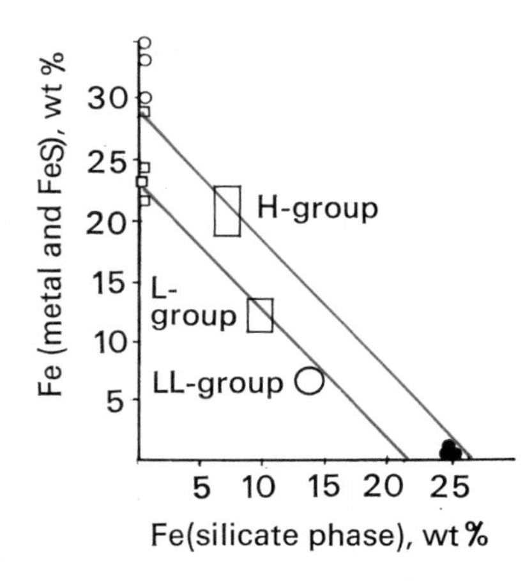

Figure 9 Relationship between metallic reduced (and sulphide) iron and oxidized (silicate) iron in chondrites. Analyses lying on the same diagonal line have the same total iron content.

SAQ 19 Arrange (1) the enstatite chondrites, (2) the carbonaceous chondrites, (3) the H Group and (4) the L Group in order of increasing *oxidation* of iron.

The enstatite chondrites are severely reduced and therefore contain very little iron in the silicate phase. The carbonaceous chondrites, in contrast, contain subordinate or no metallic iron, being the only group shown in Figure 9 in which oxidized iron significantly exceeds reduced iron. Note, too, that the low iron (LL) group of chondrites can be seen as a distinct group in this graph, since it falls below the lines of constant iron content passing through all the other chondrite groups.

According to Ringwood, the parental material of the chondrites was highly oxidized and reduction took place *during* accretion. In this way, a *range* of oxidation states could arise, depending upon varying degrees of reduction of this material, during accretion, into the different chondritic bodies. Fragmentation of these bodies could then produce the various chondrite types named in Figure 9.

The oxidation state of the Earth is believed to have been determined by the same process. The proportion of metal to silicate was thus dependent upon the degree of reduction during accretion of the Earth.

How could this reduction have taken place?

carbonaceous reducing agents

The two commonest reducing agents are carbon and hydrogen. The most reduced chondrites – the enstatite chondrites – contain various significant minerals, graphite (C), schreibersite ($(FeNi)_3P$), oldhamite (CaS), and daubreelite ($FeCr_2S_4$), as well as having some Si in solid solution in the iron phase. These minerals are found in blast furnace assemblages. They are produced by the strong reduction of iron ores by carbon. This suggests that the mineral assemblages of reduced meteorites have been produced by the reduction of oxidized materials by carbonaceous reducing agents.

Reduction by hydrogen, alternatively, would be unlikely to produce graphite in the enstatite chondrites. This is because C would escape as CO or CH_4. Similarly hydrogen would reduce troilite (FeS) according to:

$$FeS + H_2 \rightarrow Fe + H_2S.$$

The presence of about 3 per cent of sulphur in analyses of enstatite chondrites is thus explicable only on the basis that their reduction was not caused by hydrogen. The different degrees of reduction seen in different meteorite groups can be explained by the reduction, *in situ*, of oxidized meteoritic material by carbon compounds, which may have been trapped as different amounts of hydrocarbons in the accreting dust. On this basis, the enstatite chondrites once contained excess hydrocarbons, and now therefore contain graphite. This does not, however, explain the difference in Fe:Si ratio between, say, the H and L groups (Fig. 9), which must have a different origin.

The accretion of oxidized carbonaceous chondrite material preceded reduction inside the Earth, according to this model. The accreting material may have been extremely cold (< 100 K) and would then have consisted of non-volatile silicate and iron oxide dust, ice, and complex carbonaceous compounds, as well as other materials.

During the early stages of accretion, gravitational energy† would be low and would not be sufficient to retain the volatiles expelled on impact by the accreting particles. However, at a critical stage the planetary nucleus could retain some of the escaping volatiles in an atmosphere.

As the body grew and its gravity increased, small particles could be completely vaporized on impact. Larger particles might reach the solid surface and explode. This effect would result in a gradual rise of temperature at the surface (=energy of accretion). In the presence of reducing agents at these high temperatures, accreting iron oxides and silicates would be reduced.

During the latest stages of accretion, the melting point of iron would be exceeded at the surface. Metallic iron would segregate into masses. At this point the interior of the Earth would be volatile-rich and oxidized and the outer layers progressively more reduced and poor in volatiles.

This situation would not be stable. The reduced material at the surface would be denser than the inner part of the Earth. A stage would be reached at which masses of metallic iron could flow directly towards the Earth's centre. This flow would release sufficient gravitational energy to result in complete melting throughout the Earth's interior.

On this model, segregation of the core occurred as a continuous or possibly a catastrophic process during the later stages of the primary accretion. A requirement of this single stage hypothesis is the escape of a very large amount of volatiles originally combined with the oxidized materials of the primitive Earth. These may have amounted to half the mass of the Earth and would have consisted chiefly of CO and H_2 with elements that were volatile under the conditions of the early accretion. Evidence that the present atmosphere is secondary was presented in S100[11], and it is also a feature of Hypothesis I.

SAQ 20 There are two plausible theories for the origin of the Earth's differentiation into core and mantle. These are (I) a multi-stage origin and (II) a single-stage origin. Match the following statements to these two theories.

1 The accreting material consisted of:

(a) oxides and silicates as in C1 chondrites;

(b) oxides, silicates and metallic iron particles as in ordinary chondrites.

2 *Initial* heating was mainly accomplished by:

(a) accretional energy;

(b) radioactive decay.

3 Metallic iron first melted:

(a) at surface;

(b) in a zone at a depth of several hundred km.

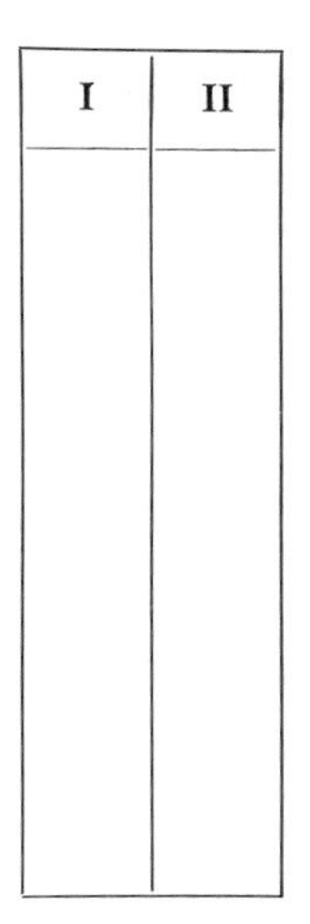

Following separation of the homogeneous Earth into core and mantle, the next major chemical event was the development of a crust. This is likely to have originated from igneous rocks derived by partial melting within the upper mantle.

Can you recall an igneous rock formed from partial melting of upper mantle peridotite?

A common rock resulting from partial melting of mantle peridotite is basalt. Other igneous rocks may, however, have been present in the earliest formed crust. Note that for the separation of the crust, mantle, and core it is not necsssary for the Earth to have been liquid throughout. Smaller degrees of partial or total melting are completely adequate.

3.8.2 Chemical partition of elements during the formation of the core, mantle, and crust

chemical partition of elements between core and mantle

Read pp. 54–5 in Principles of Geochemistry (The primary differentiation of the elements.) Note the following points:

1 Meteorites and the Earth consist essentially of Fe, Mg, Si, O, and S. The oxygen is insufficient to completely oxidize the electropositive elements† Fe, Mg and Si.

2 Under the conditions of early differentiation of the Earth, a system of three immiscible phases may have been formed. These would be Mg-Fe silicate, iron sulphide, and metallic iron.

> **SAQ 21** Why are these phases immiscible?

3 These phases would separate *according to their relative densities* and the distribution of the remaining electropositive elements would be governed by reactions of the type below.

M+Fe silicate $\rightleftharpoons$ M silicate+Fe........................(1)
M+Fe sulphide $\rightleftharpoons$ M sulphide+Fe ....(2)
Fe silicate+M sulphide $\rightleftharpoons$ M silicate+Fe sulphide................(3)

M means any cation other than Fe.

> **SAQ 22** In the case of one particular element (M) participating in these reactions, reactions (1) and (3) move to the right. Would you expect this element to become concentrated into core or mantle phases?

We can predict the behaviour of elements in reactions (1) to (3) above, if we know whether they are more or less readily oxidized than iron (cf. Section 3.3). Clearly, elements more easily oxidized than iron would be more likely to enter mantle (silicate) phases than those less easily oxidized, which would tend to enter the core (metallic) phase. A qualitative order of 'ease of oxidation' is shown in Table 7.

Table 7 Qualitative order of oxidation for some important elements. The elements are expressed as stable oxides.

relative ease of oxidation of elements

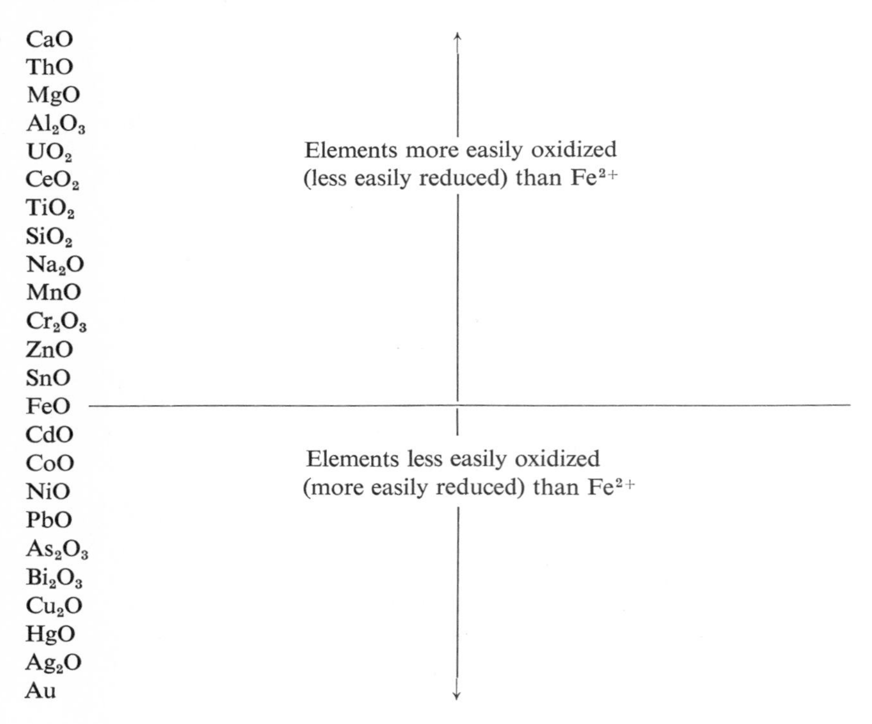

This recapitulates, in a slightly different way, our conclusions from Section 3.3, about how elements may be distributed throughout the Earth as a whole and how elements may occur at the Earth's surface. Considering the Earth *as a whole*, the elements more easily oxidized than Fe^{2+} are concentrated in the mantle and crust; those elements less easily oxidized than Fe^{2+} will be partly reduced and concentrate in the Earth's core. Such elements will include Ni and Co.

At the Earth's surface, in a general way, the elements more easily oxidized than Fe^{2+} will tend to occur as oxides and silicates; those less easily oxidized will tend to occur in reduced forms – as sulphides or as the native metal.

Iron is used as a datum since it is assumed to occur in reduced form in the core and in an oxidized form in the mantle and crust. These general features are confirmed by a study of meteorites where Fe, Ni and Co are concentrated into the metal phase, with SiO_2 and MgO making up the bulk of the silicate phase (Table 1).

SAQ 23 Meteorites vary in their overall degree of oxidation. In one of the most reduced groups, some of the titanium occurs in a reduced state, as a titanium sulphide. Look at Table 7. Under these conditions would you expect to find some (a) Al, (b) Cr, (c) Ni, in a reduced state (i.e. as sulphide or native metal)?

This picture is oversimplified. Although the crust and mantle do consist overwhelmingly of the elements predicted from above the line in the list given in Table 7, small amounts of all the other elements do occur, although almost all of them are in reduced form as metal or sulphide. Similarly the core, although believed to consist largely of Fe-Ni, has been suggested to contain small amounts of some more easily oxidized elements such as Na, Mg, and Si (Unit 2, Section 2.15) in reduced form.

SAQ 24 Analyses of gold in chondrites show concentrations of about 1.5 ppm in the metal phase and about 0.005 ppm in the silicate phase (Table 1). Read the notes below, and then calculate the weight of gold in the Earth's mantle and core, and hence the proportion of the Earth's gold that is in the core only.

Notes

1 This involves making the assumption that the relative abundance of gold in the silicate and metal phases of chondrites is the same as in the mantle and core of the earth.

2 For the purposes of the calculation assume the mantle to have a mass of 4×10^{27}g and the core a mass of 2×10^{27}g.

Composition of the core

We can study the mode of occurrence of elements within the crust and within samples of the upper mantle. We know that the common rock types are composed very largely of silicates with small amounts of oxides. We know rather less about the chemical features of the core! However, having established some models for the formation of the core, let us look at possible details of its chemical composition.

From Table 7, after iron, which *abundant* element is most easily reduced?

Under reducing conditions more severe than required to reduce iron, silica will be reduced to elemental silicon. In Unit 2 it was suggested that some silicon may be alloyed in the Earth's core. This is a popular choice for a constituent of the core. The presence of silicon in the Earth's core is consistent with the reduction of silicates at the Earth's surface (Hypothesis II) followed by solution of the silicon in the metallic iron. Although silicon is a popular choice for a component of the core, not all Earth scientists agree that it is the best one.

Another possibility is sulphur (p. 51 in *Principles of Geochemistry*). When the formation of the core is considered, there are several reasons which appear to favour sulphur rather than silicon. One reason is that an Fe-S melt needs much less drastic conditions for its formation than does an Fe-Si melt, which requires high temperature reduction of silicates. An Fe-S melt would form at only

990 °C within the Earth whereas Fe-Si melts would form at temperatures only slightly less than the melting point of iron, i.e. above about 1 500 °C (Fig. 3.10 in *Principles of Geochemistry*). Movement of Fe-S to the centre of the Earth may thus be easier than that of Fe-Si alloys. For these reasons, some scientists consider a Fe-Ni-S alloy more likely than Fe-Ni-Si for the major element composition of the Earth's core.

Although the nature of the light-alloying material in the core is not agreed upon, it is widely accepted that the core consists chiefly of iron and nickel with smaller amounts of other siderophile elements.

The formation of the Earth's core was thus accompanied by a partitioning of elements. Some elements were capable of being reduced and of alloying with iron, the major element of the core – and so they also became concentrated into the core. Elements which were not reduced formed the mantle and primitive crust. This major twofold division of the Earth does not account for the third phase which may have been present within the Earth, the sulphide phase.

We have compared the present Earth with ordinary chondrite meteorites. As you know these contain metal, sulphide and silicate phases. We think the Earth contains metal and silicate phases but does it contain a sulphide phase? How we answer this depends upon what we think is present in the core. If an Fe-S melt formed at an early stage, this may now be present within the core. However, Fe-Ni sulphides are less dense than pure metals and may have a similar density to parts of the lower mantle. If early separation of an Fe-S melt did not occur, the Earth's sulphides may be concentrated in this region of the Earth. A final alternative is that the Earth's sulphides are dispersed throughout core, mantle and crust, for sulphides are common minor minerals in many crustal rocks.

primary geochemical differentiation

The partitioning described above is often termed the *primary geochemical differentiation*. During this differentiation the distribution of the elements was controlled by their affinities for the major phases that formed (Section 3.2). This affinity is, in turn, controlled by the electronic configurations of their atoms (Section 3.3). The distribution of the elements is controlled by gravity acting upon contrasting densities of the major phases. As emphasized very clearly on p. 55 in *Principles of Geochemistry*, it is *not* controlled by atomic weight.

secondary geochemical differentiation

The processes taking place in and on the crust are known as *secondary geochemical differentiation*. During this later geochemical evolution of the Earth a greater variety of processes assumed importance. These include the formation of an early atmosphere and the organic reactions which initiated the development of life (S100[19]). Both primary and secondary differentiation processes are summarized in the flow diagram, Figure 10.

3.9 Summary

We suggest you read this summary and then look back at the conceptual diagram at the beginning of the Unit.

1 By analogy with meteorites and metal smelting furnaces, elements may be classified according to their affinity for the metal, sulphide, silicate or gaseous phase. These are respectively termed siderophile, chalcophile, lithophile and atmophile elements.

2 The partitioning of elements within a body such as the Earth depends upon the overall chemical composition of the system, its degree of oxidation and the atomic structure of the elements concerned.

3 The chemical composition of the Sun is essentially that of the Solar System, since the Sun contains over 99 per cent of the mass of the Solar System. Element abundances in the Sun provide the best estimate for the abundance of volatile elements in the Solar System.

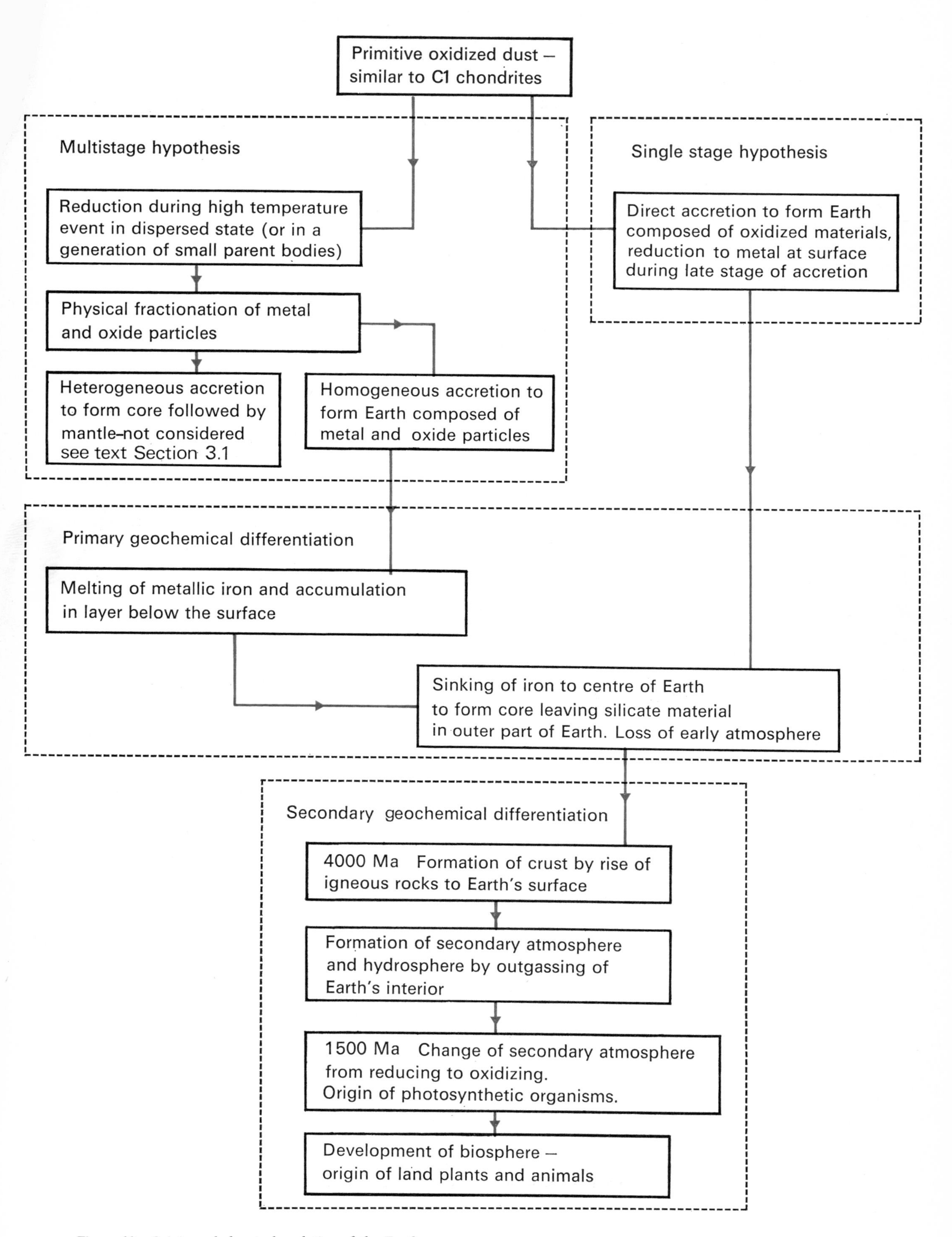

Figure 10 Origin and chemical evolution of the Earth.

4 Meteorites are solid samples of the non-volatile matter of the Solar System. Of the several varieties, the carbonaceous chondrites, Group 1 (the C1 chondrites) are considered to provide the best estimate of the primitive non-volatile matter of the Solar System.

5 Rocks exposed at the Moon's surface cannot be representative of the Moon as a whole; they can therefore not yet tell us much about the origin of the Moon or of the Solar System.

6 The development of the Earth took place from a homogeneous chondritic body. This may have originated by the accretion of material like ordinary chondrites, with core formation being initiated by heat from naturally radioactive elements. Alternatively, the accreting material could have been of C1 chondrite type, in which case the energy of accretion may have initiated the formation of the core.

7 During the formation of the core (termed the primary geochemical differentiation), the elements became partitioned between metal, sulphide and silicate phases. This was followed by the secondary geochemical differentiation – the formation of the crust and of the hydrosphere, biosphere and atmosphere.

Further Reading (Black Page)

Wood, J. A. (1970), *Meteorites and the Origin of Planets*, McGraw-Hill.

Ahrens, L. H. (1965), *Distribution of the Elements in our Planet*, McGraw-Hill.

Appendix 1

References to S100

1 Unit 22, *The Earth: Its Shape, Internal Structure and Composition*, Section 22.5.
2 Unit 27, *Earth History*, Section 27.1.6.
3 Unit 8, *The Periodic Table and Chemical Bonding*, Table 6.
4 Unit 8, Section 8.3.
5 Unit 8, Appendix 3.
6 Unit 6, *Atoms, Elements and Isotopes: Atomic Structure*, Section 6.5.6.
7 Unit 6, Section 6.5.5.
8 Unit 8, Appendix 1.
9 Unit 8, Section 8.2.
10 Unit 8, Section 8.5.1.
11 Unit 27, Sections 27.1.4, 27.3.1.
12 Unit 22, Section 22.2.
13 Unit 27, Section 27.1.
14 Unit 6, Section 6.4.6.
15 Unit 31, *The Nucleus of the Atom*, Section 31.5.3.
16 Unit 27, Section 27.1.2.
17 Unit 6, Section 6.2.1.
18 Unit 6, Section 6.2.2.
19 Unit 27, Section 27.2.

Appendix 2

Glossary

CHLORITE — A hydrated layer silicate mineral characteristically green in colour. Chlorites show extensive solid solution and have the general formula $(MgFeAl)_{12}(AlSi)_8O_{20}(OH)_{16}$. Chlorites are common in metamorphic rocks, formed at relatively low temperatures on Earth (pp. 158–60 in *Principles of Geochemistry*).

ELECTROPOSITIVE ELEMENT — An element which gives up valence electrons readily, to form a positively charged ion.

GRAVITATIONAL ENERGY — The energy liberated from a system as it changes from a gravitationally unstable to a more stable state.

ILMENITE — An iron-titanium oxide ($FeTiO_3$) common as a minor constituent of terrestrial basaltic rocks (Unit 2, Section 2.10) – also found in lunar basalts.

Self-assessment Answers and Comments

SAQ 1 1, B; 2, A; 3, C; 4, A; 5, B; 6, C; 7, A; 8, C.

Copper and lead, although dominantly chalcophile, are found in both the silicate and metal phases. Calcium, magnesium and the trace element, thorium, are very strongly lithophile and occur in extremely low concentrations in metal and sulphide phases. Gold and platinum show a strong siderophile tendency, much stronger than, say, nickel or iron. The alkali metals (Li, Na, K, Rb, Cs) are so strongly lithophile that they are virtually undetectable in sulphide and metal samples.

SAQ 2 (a) A, ii; B, i; C, iii.

(b) A, 3; B, 1; C, 2.

The elements are A, nickel; B, caesium; C, platinum.

SAQ 3 1, A; 2, B; 3, D; 4, A; 5, C; 6, D; 7, B.

Answers to 3 and 6 include some less easily classified elements (the most important being Fe) which occur in a variety of ways, including the native element.

SAQ 4 The Solar spectrum is essentially composed of (a) and (c).

SAQ 5 These are the C and F lines in Figure 1, corresponding to hydrogen – the major component of all known stars.

SAQ 6 1, 3, 4 are correct. Answer 2 is impossible – the chemical composition of a star cannot be determined unless operations 1, 3 and 4 have been carried out!

SAQ 7 1 B, D, G.
2 B, E, F.
3 B, C.
4 A.

SAQ 8 (a) The group of meteorites most frequently recovered on the Earth's surface is the irons. This is because they are so distinct in their physical features that they can easily be recognized as meteorites when found. The stony meteorites can easily be overlooked, unless actually seen to fall. As you can see from Table 2.5. in *Principles of Geochemistry*, stony meteorites are much more abundant than irons.

(b) The abundances of O and Fe and of S and Ni are reversed (cf. Table 3.9 in *Principles of Geochemistry*).

SAQ 9 (a) Because iron meteorites are not so homogeneous as the chondrites, larger samples of irons are needed to obtain an accurate estimate of the troilite (and hence S) content. A reasonably good sampling of the troilite of an iron meteorite may require the analysis of over 10 kg of material. Since few analyses use samples as large as this (and many irons weigh less than this anyway) the reports of sulphur abundances in iron meteorites are often not meaningful and are notoriously unreliable.

(b) Because chondrites are more homogeneous, smaller samples can be used and S abundances are more reliable.
These considerations also apply to the *chalcophile* elements which will be found in the sulphide phase.

SAQ 10 The types and names of the meteorites are given below:

A Tieschitz chondrite. The section shows chondrules and grains of olivine and pyroxene (hypersthene*) in a dark matrix.

B Stannern achondrite. You will see a larger fragment of this in TV programme 3. This section shows dark pyroxene (augite*) between light laths of calcium-rich plagioclase.

C Shergotty achondrite. You can see dark pyroxene (augite*) and patches of lighter material which, in this case, is glass with the chemical composition of plagioclase.

D Indarch chondrite. This contains chondrules and grains of pyroxene (enstatite*) in a black matrix of nickel-iron and sulphides.

SAQ 11 Carbon (C) is approximately 10^6 times (six orders of magnitude) more abundant than gallium (Ga). This can be explained by the fact that the atomic number of C is *low* and *even*, whereas that of Ga is *high* and *odd*.

SAQ 12 (a) True; (b) False.

SAQ 13 The order is Cr, Rb, Cl, Hg.

SAQ 14 1, False; 2, True; 3, False; 4, True; 5, True; 6, True, 7, True.

SAQ 15 Only 2 is possible. 1 is wrong because below about 300 km basalt changes into eclogite, which is much denser than the Moon as a whole. 3 is wrong because the density of lunar basalt is almost exactly the same as that of the Moon as a whole.

SAQ 16 Under the pressure conditions of the lunar interior known lunar rocks would transform to denser types, giving a lunar density too high to agree with independent density measurements.

SAQ 17 The half lifes of ^{40}K and ^{235}U are short in comparison with ^{238}U and ^{232}Th. The former two elements have therefore been destroyed by radioactive decay at a greater rate than the latter two.

SAQ 18 (a) Yes, 600 Ma ago the melting point of pure iron (1 800°C) was exceeded, *but only in a comparatively narrow zone at a depth of about 400 km* inside the Earth.
(b) No, at this and subsequent times the temperature was well below the melting point of diopside at any depth; since diopside is related to augite, a major constituent of basalt, its melting temperature is only exceeded when basalt is formed (actually that is an oversimplification, as you will see in Unit 4).

SAQ 19 The order is 1, 3, 4, 2.

SAQ 20 1 (a) II 1 (b) I
2 (a) II 2 (b) I
3 (a) II 3 (b) I

SAQ 21 The three phases are immiscible because of their different atomic structures. In iron the bonding is metallic, in silicates 50–80 per cent ionic, and in sulphides 80–90 per cent covalent. This ensures that the three phases have different physical properties and will segregate rapidly as soon as they become hot enough to approach the liquid phase. An analogous mixture would be one containing oil, water and mercury.

SAQ 22 The element would become concentrated into mantle phases, since it tends to form silicates.

* *Hypersthene* ($(FeMg)_2Si_2O_6$), *augite* ($Ca(Fe,Mg)Si_2O_6$) *and enstatite* ($Mg_2Si_2O_6$) *are all varieties of pyroxene* (*Unit 2, Section 2.7.4*).

AQ 23 If conditions are so reducing that Ti occurs as sulphide, then Ni and Cr would certainly occur in a reduced state. Ni occurs in the metallic phase in all groups of meteorites and in the most reduced groups Cr shows chalcophilic tendencies and occurs in daubreelite ($FeCr_2S_4$). About Al we could be less certain, but, in fact, it is lithophile in all meteorite groups.

SAQ 24 We assume that:

Gold in mantle = 0.005 ppm.
Gold in core = 1.5 ppm.

$$\text{Weight of gold in mantle} = \frac{0.005}{10^6} \times 4 \times 10^{27} = 2 \times 10^{19}\,\text{g}.$$

$$\text{Weight of gold in core} = \frac{1.5}{10^6} \times 2 \times 10^{27} = 3 \times 10^{21}\,\text{g}.$$

Total weight of gold in the Earth = 3.02×10^{21} g.
Proportion of Earth's gold which is in the core

$$= 3 \times 10^{21} / 3.02 \times 10^{21}$$
$$= 3/3.02 = 99.3 \text{ per cent.}$$

Acknowledgements

Grateful acknowledgement is made to the following for illustrations used in this Unit:

Figures 1 and 2: Van Nostrand Reinhold for H. E. White, *Modern College Physics;* Figure 3: Lick Observatory for G. Abel, *Exploration of the Universe;* Figure 4: University of Chicago Press for three photographs from J. A. Wood, *The Moon, Meterorites and Comets* and Maxwell Scientific International Inc. for one photograph from B. Mason, *Geochimica et Cosmochimica Acta*, **30**, 1966; Figure 5: University of Chicago Press for J. A. Wood in *The Moon, Meteorites and Comets;* Figure 6: Pergamon Press Ltd, for E. Anders *Geochimica et Cosmochimica Acta*, **35**, (5), 1971; Figure 7: Artemis Press for B. Mason, *Understanding the Earth;* Figure 8: Maxwell Scientific International Inc. for E. R. DuFresne and E. Anders, *Geochimica et Cosmochimica Acta*, **26**, 1962; Figure 9: Springer Verlag for K. Keil, 'Classification of chondrites' in *Handbook of Geochemistry*, K. H. Wederpohl (ed.).

GEOCHEMISTRY